The Geology of Liberia: a Selected Bibliography of Liberian Geology, Geography and Earth Science.

By R. Lee Hadden

Originally prepared by the US Geological Survey Library staff as part of an
US Department of State project to restore the Geological Library of Liberia, 1998-1999.
Revised and Updated through 2006.

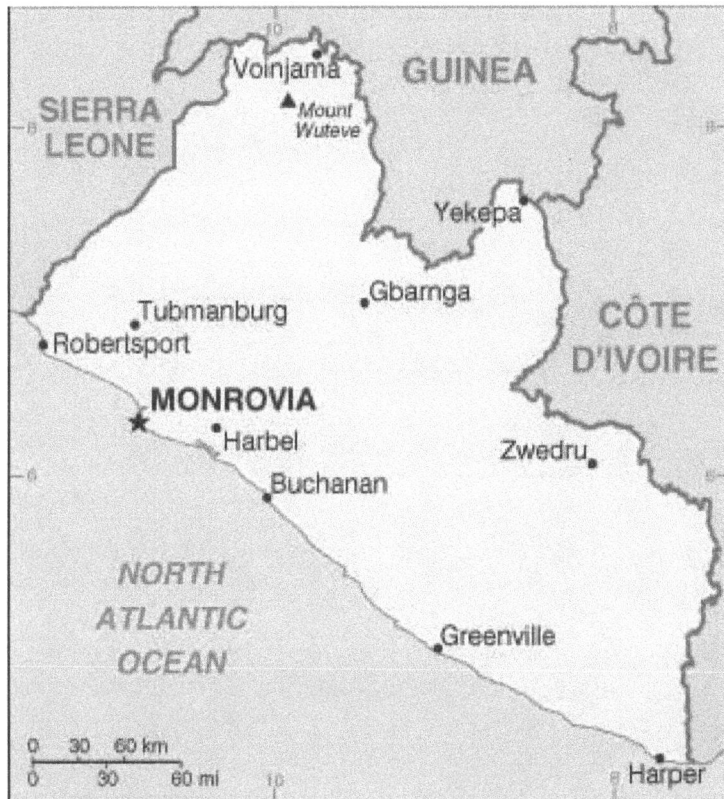

Topographic Engineering Center   May 2006
US Army Corps of Engineers
7701 Telegraph Road
Alexandria, Virginia 22315

UNCLASSIFIED / UNLIMITED

Preface

Government representatives of the Republic of Liberia and the US Department of State visited the US Geological Survey Library in late 1998 and again early in 1999. The National Library of Liberia in Monrovia was destroyed by the awful civil war in their country, and their representatives desired assistance to rebuild their country's written heritage. At the request of the Department of State, the USGS Library was able to supply them with a complete bibliography of USGS publications written about Liberia.

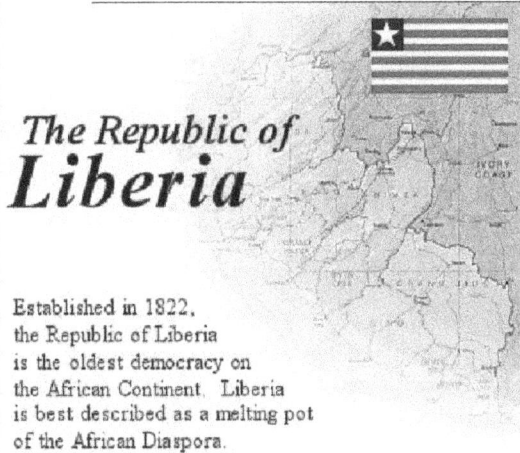

The Republic of

Liberia

Established in 1822, the Republic of Liberia is the oldest democracy on the African Continent. Liberia is best described as a melting pot of the African Diaspora.

The USGS Library staff was also able to supply them with copies of other publications written about their country by their own Liberian Geological Survey, as well as reports and maps by other private, commercial and governmental organizations. It was discovered that the USGS Library had a far greater collection on the geology of Liberia than their national library had owned before it was destroyed.

The Liberian representatives applied for a grant from the US Department of State to cover the cost to copy the materials published by the US Geological Survey in the USGS Library for the use of their library, since these government documents and maps had no copyright restrictions. Later, the plan was to extend the bibliography to comprehensively cover the geological literature of Liberia, as they intended to purchase those copyrighted items needed to refill their library collections.

By collecting so many publications over so many years through exchanges, gifts and purchases, the US Geological Survey Library staff has built up a significant collection of the world's knowledge of the earth's sciences. It was also a remarkably cost effective collection, since many of these publications were acquired through scientific exchange with the government of Liberia and their agencies. As a result, there was little cash spent on obtaining their reports and maps, since for many years, extra copies of USGS scientific publications were simply exchanged for their geological publications. By collecting in this manner over time, the USGS Library acquired an excellent reservoir of earth science information about their country, which cannot be compared with any other library for the breath and scope of our collection of Liberian materials.

By doing this with so many different countries, of which Liberia is only one example, and exchanging their reports, books and maps and reports with their private organizations as well, the USGS Library has been able to build up a massive amount of information that is freely available to the earth science community, at relatively little cost.

The willingness to share the USGS Library collection and services with Liberia and other countries whose libraries are harmed by natural disaster or war, recapitulates the USGS Library mandate for the increase and diffusion of knowledge.

The backbone of this bibliography are the publications of the US Geological Survey. Without a doubt, the largest number of publications have been written by the US Geological Survey, often in cooperation with the Liberian Geological Survey. In this revised and updated report, additional materials have been identified and located in university, government and corporate libraries around the world.

Wherever possible, the information concerning the location of the identified materials or where they can be viewed is also given. Standard numbers, such as ISSNs, ISBNs and OCLC numbers are listed when possible to help locate and verify these publications. While materials in some libraries, particularly maps, frequently can't be borrowed, others can be sent through inter-library loan agreements or by other library collaborations.

Other materials can be located and purchased through various document delivery companies such as AMRS or the American Geological Institute. For a fee, they will locate and supply to their customers copies of journal articles, maps or reports.

Introduction

The Republic of Liberia has it's origins in 1821, when the American Colonization Society began it's campaign to send freed slaves from the United States back to Africa. The country was settled by freed American slaves from 1822 through the 1860s, who had few interactions with the indigenous people. The republic, dating from 1847, is the oldest independent country in Africa. Liberia, meaning "Land of the Free", has about 43,000 square miles, and is about the size of the state of Tennessee.

Liberia is a few degrees north of the equator and lies along the great western bulge of the continent. The coastline of Liberia is nearly 370 miles long, and extends from West Africa westward to Sierra Leone. Going inland, the republic varies from 100 to 200 miles in width, and ascends from sea level to the Guinea Highlands and the country of Guinea. The "Pepper Coast" is the name of a coastal area in western Africa, between Cape Mesurado and Cape Palmas. It encloses the present republic of Liberia and got its name from the melegueta pepper. It is also known as the "Grain Coast."

Liberia is divided into 15 counties and they are subdivided into districts. The counties include: Bomi; Bong; Gbarpolu; Grand Bassa; Grand Cape Mount; Grand Gedeh; Grand Kru; Lofa; Margibi; Maryland; Montserrado; Nimba; River Cess; River Gee and Sinoe.

Regretfully, Liberia has witnessed two civil wars, the Liberian Civil War (1989–1996), and the Second Liberian Civil War (1999–2003), that have displaced hundreds of thousands of people and destroyed their economy and the nation's infrastructure.

Transportation in 1999 had a total road network of about 10,600km. There were 657km of paved roads and 9,943km of unpaved roads throughout Liberia.

Geology

Liberia is perched on the West African Shield, a rock formation from 2.7 to 3.4 billion years old. The West African Shield that is made of granite, schist, and gneiss. In Liberia this shield has been intensely folded and faulted and is interspersed with iron-bearing formations known as itabirites. Along the coast lie beds of sandstone, with occasional crystalline-rock outcrops.

Monrovia stands on such an outcropping, a ridge of diabase (a dark-colored, fine-grained rock). Most of the crystalline rocks are of Precambrian age. The western half of country is typically of Archean age. In the eastern half of the country, lenses of Proterozoic greenstone belts occur surrounded by rocks of probable Archean age. Rocks of Pan African age extend northwesterly along most of the Liberian coastline from the Cestos shear zone.

Along the Atlantic Ocean, the coastline is characterized by lagoons, mangrove swamps, and river-deposited sandbars. Inland, the grassy plateau supports limited agriculture.

Climate

Liberia is known for its sustained heat and heavy rainfall.

Because the republic lies south of the Tropic of Cancer and only a few degrees north of the equator, the days vary little in length. The tropical solar radiation is intense and the radiation is uniform across the country. Temperatures remain warm throughout the country, and there is little change in temperature between seasons. The mean annual temperatures in Fahrenheit range from the 70s to the 80s. The mean monthly maxima decline from the low 90s to the mean 80s during the rainy season. The mean monthly minima range from the low 60s in the highlands of the northwest to the low 70s at Monrovia and along the coast. Temperatures inland are warmer than along the coast, but the diurnal range is also greater inland.

More rain falls than in other areas of West Africa. The relative humidity is high throughout the county, and averages from 70 to 90 per cent, especially along the coat. The continental and maritime masses of air alternate their movements back and forth, and from north to south. This brings some seasonal differences in rainfall intensity. The coastal region has the heaviest rainfall, from between 155 to 175 inches annually in the west, and with nearly 100 inches of rain annually in the southeastern part of the country. Monrovia receives almost 180 inches of rain annually.

Rainfall decreases going north and inland. But the rainfall increases again in the highlands and the northernmost part of the republic. The driest part of the country is along a strip of the eastward flowing Cavalla River, but even there, the land receives over 70 inches of rain a year.

In Liberia, the rainy season begins in April or May, and reaches a peak in July through September, and tapers off again in October. Monrovia and Buchanan, on the coastal plains, receive a heavy rain earlier in the season, then they experience a period of reduced rainfall called the "middle dries" before heavy rains return in August. In the southeastern part of the country, the rainy season begins in April and lasts for two or three months, and then is followed by a drier period of two or three months. Then a second rainy season begins in September and lasts until November. The "middle dries" are not dry enough to be called a true dry season.

Water supplies have been improved in both rural and urban areas so that some 40 percent of the population has access to potable water. Surface water is abundant, and groundwater reserves are ample and regularly replenished by the country's heavy rainfall.

Rivers

The major rivers of Liberia are the Cavalla, the Cestos, the Lofa, the Mano, the Morro, the Saint John and the Saint Paul. The Mano and Morro rivers in the northwest and the Cavalla River in the southeast are boundary lines for part of the country. Most of the rivers of Liberia flow from the mountains inland in the northeast to the coast in the southeast, and parallel each other. Among the low mountains and hills, the river beds are steep and irregular, with frequent falls or rapids. Many rocks, waterfalls, rapids and sandbanks reduce navigation of these rivers very far inland. Closer to the coast, the river grade becomes less, and tidal current prevent the rivers from removing sand bars and accumulations. However, most streams overflow their banks regularly, and during the rainy seasons there is often severe flooding along the coastal plains. Many rivers flow long the coast for miles before they enter the Atlantic Ocean.

The rivers have been harnessed to generate hydroelectric power. The Farmington River is one source of hydroelectric power. The Mount Coffee hydroelectric station outside Monrovia on the St. Paul River is the country's largest hydroelectric installation. Electrical production in Liberia from all sources was 509.4 million kWh in 2003.

The Cavalla River in western Africa runs between the Ivory Coast and Liberia. The river is alternately known as the Cavally, Youbou, or Diougou River. The Cavalla rises north of the Nimba Range in Guinea and flows south to form more than half of the Liberia and Côte d'Ivoire border. The Cavalla enters the Gulf of Guinea 13 miles (21 km) east of Harper, Liberia, after a course of some 320 miles (515 km). With its major tributaries, the Duobe and the Hana, it drains an area of 11,670 square miles (30,225 square km).

The St. Paul River was first sighted by Portuguese sailors in the 15th century on St. Paul's Day. The river begins in southeastern Guinea, crosses into northern Liberia about 30 miles (50 km) due north of Gbarnga, in Bong County. It then flows through Montserrado County, and eventually becomes the dividing line between Monrovia and Brewerville where it flows into the Atlantic Ocean.

Topology

The main physiographic regions of Liberia parallel the coast. These regions are: the coastal plains, the rolling hills, and the highlands. The Forest Zone covers all of Liberia.

The coastal plains are about 350 miles (560 kilometers) long and extend up to 25 miles inland. They are low and sandy, with miles of beaches interspersed with bar-enclosed lagoons, mangrove swamps, and a few rocky promontories. The highest promontory is Cape Mount (about 1,000 feet or 305 meters in elevation) in the northwest, with Cape Mesurado in Monrovia, and Cape Palmas in the southeast. Its deepest extensions lie along the watercourses. The shore is broken by river estuaries, tidal creeks, swamps, and a few rocky capes and promontories that appear as landmarks from the sea. Except for those promontories and capes and an occasional small hill, the altitude of the coastal region usually rises no higher than 30 to 60 feet. The mouths of the rivers are so obstructed by shifting sandbars and submerged rocks that there are no natural harbors. The surf is normally heavy all along the coast but is worse at the height of the rainy season.

Parallel to the coastal plains is a region of rolling hills some 20 miles wide with an average maximum elevation of about 300 feet; although a few hills rise as high as 500 feet. It is a region suitable for agriculture and forestry. Further on, the country consists of rolling plateaus and low-lying hills rising to the higher elevations of 600 to 1,000 feet that constitute almost half of Liberia's terrain. In the far northwest and north central portions of the territory are the outliers of the Guinea Highlands. This land is well watered, and a number of narrow, roughly parallel river basins run to the sea at right angles to the northwest-southwest trend of the belts of relief.

Most of this country lies in the heaviest rainfall zone in West Africa. Precipitation, however, decreases progressively inland, and rainfall belts, like relief belts, run roughly parallel to the coast. There is normally some rain during every month of the year, but most of the country is characterized by wet and dry seasons. The climate is warm and humid, and the annual temperature variation is quite small. At the northern edge of this belt, a steep rise indicates the southern edge of a range of low mountains and a plateau that constitutes nearly half the country's interior.

The highlands are behind the rolling hills, most of the country's interior is a dissected plateau with scattered low mountains ranging from 600 to 1,000 feet in elevation. The long ridges and dome shaped hills that constitute the northern highlands are part of the Guinea Highlands and occupy those sections of Lofa and Nimba counties that thrust much farther north than the rest of

Liberia's boundary with Guinea and Ivory Coast. These mountains, mainly the Wologizi Range in Lofa County and the Nimba Range north of the town of Sanniquellie, rise to altitudes above 4,000 feet. Mount Wutivi, the highest peak in the Wologizi Range, reaches about 4,450 feet, and the Nimba Range's Guest House Hill is, at 4,540 feet, the highest point in Liberia.

In West Africa, the forest zone refers to the southern part of the region once largely covered by tropical rainforest. The forest zone of West Africa, in the strict sense, covers all of Liberia and Sierra Leone, most of Guinea, the southern halves of Côte d'Ivoire and Nigeria, and parts of Ghana, Togo and Guinea-Bissau. In the eastern part of the forest zone, because of the influence of Mount Cameroon, soils are often fertile and there are large areas of subsistence farming. Major crops include millet, yams and rice, whilst plantation agriculture is extensive on the best soils, producing chiefly cocoa. Further west, due to the ancient geology of the region, soils are much less fertile and farming becomes chiefly confined to the raising of perennial crops, with cocoa remaining pre-eminent. Forestry has devastated much of the natural rainforest in countries such as Côte d'Ivoire and Liberia. Farmers without land have been pushed onto land with marginal soil for agriculture by population growth, which, despite frequent warfare, continues to be among the highest in the world.

Land Resources

Soil- More than 80 percent of Liberia's soils can be used for agriculture. Although there have been some local soil surveys, countrywide data are insufficient for a broad evaluation of soil potentials and agriculture. This is an area for future research. Cultivable land to meet the needs of the subsistence population, as well as for expansion of export tree crops, was quite satisfactory.

According to estimates of the Food and Agriculture Organization (FAO) in the early 1980s, only about 1,430 square miles of the country's total land area (roughly 3.9 percent) were used for cultivation. Permanent tree crops, such as rubber, coffee, and cacao, occupied 946 square miles, or two-thirds of the cultivated area; short-life crops, mainly foods, were produced on about 485 square miles. The FAO also calculated that more than 21,000 square miles of additional land was in a temporary bush and tree fallow state, and much of this is at a stage available for agricultural use. There was little pressure on the fallow areas in the less heavily populated rural regions, and about 80 percent of the subsistence farmers in those regions were reportedly using for crops new land on which the age of the tree or bush stands was seven or more years. The situation was different, however, in heavily peopled areas near the towns where the fallow cycle on good land has been found to be as short as four years, a period generally inadequate to allow the replacement of natural soil nutrients.

Four types of soil are found in Liberia: latosols, lithosols, regosols and alluvial soils.

Latosols are of low to medium fertility and occur in the rolling hill country and cover about 75 percent of the total land surface in Liberia. Latosols, (formed from "laterite" and "solum", which is Latin for soil), are a soil that is rich in iron, alumina, or silica, and which formed in tropical woodlands under very humid climate with relatively high temperature. These latosols were formed on the extremely old, largely granites gneisses and other gneissic and schistic bedrock that underlie most of the country. These soils have been intensively leached by the heavy tropical rainfall and are of only medium to low fertility. Latosols are the soils on which upland rice, the largest single food crop in Liberia, is grown. Their limited amount of plant nutrients requires, without the use of fertilizer, a constant shifting of cultivation to new fields in order to maintain subsistence production levels. Large areas of these soils also support the country's major tree crops.

Shallow and coarse lithosols, in the hilly and rugged terrain, cover about 16 to 17 percent of the land in Liberia. Lithosols are a thin soil consisting of rock fragments, and is a soil with poorly defined layer horizons that consists mainly of partially weathered rock fragments. These are soils that are characterized by imperfect weathering and have low humus and mineral nutrient content. Although they support tree and other woody vegetation, these soils have little value for agriculture.

Infertile regosols, or sandy soils, are found along Liberia's coastal plains. Regosols cover about 2 percent of Liberia, and are found along the coast that is generally infertile, although they support large numbers of coconut trees, as well as oil palms. Regosols are a type of soil consisting of unconsolidated material from freshly deposited alluvium or sand.

Highly fertile alluvial soils represent only about 3 percent of the land area of Liberia, and these soils are utilized largely for agriculture. Alluvial soils are found in the river bottoms, and in swamp soils. Swamp soils, especially those known as half bog soils, are naturally rich in humus, and when drained they provide excellent conditions for swamp rice and similar crops.

The principal food crops grown are rice, mostly of the upland variety, and cassava. These crops were grown throughout the country in the traditional sector, but cassava cultivation was more heavily relied on the southeastern coastal region, where rainfall and cloud conditions were less favorable for rice. A variety of vegetables were also grown to supplement the two main staples. Climate and soils in Liberia were variously well suited to tropical tree crops, including rubber, coffee, cacao, oil palm products, and coconuts. Tree crops have been a major source of export earnings; in the period between 1979-81 rubber, coffee, cacao, and oil palm products accounted for almost one quarter of all export receipts.

With the exception of a small area in the northwest bordering Sierra Leone, the narrow coastal zone, and a region in the southeast, all of Liberia was considered ecologically suitable for commercial production of rubber. The area potentially usable for coffee cultivation was also large. In general, cacao could be grown throughout the same area; but soils required for satisfactory tree growth were less extensive, and rainfall factors placed some restrictions on profitable commercial cultivation. Oil palms grew naturally and were widely distributed, but for commercial planting the southeastern one third of the country offered the greatest future possibilities.

Mining
Gold

Gold in Liberia is mined almost entirely from alluvial deposits. Gold mining began in 1881 with the establishment of a Liberian-owned company. Other operators and individual miners exploited gold-bearing alluvial deposits in the early 1900s, but the total amount of gold recovered before War World I was routinely quite small. After WWI, gold was found in numerous river and stream deposits throughout Liberia, and placer mining became widespread. Mine output varied greatly, and many deposits were small and they were soon exhausted. In 1938 some 2,080 ounces were exported. In 1943 a new discovery of gold in Grand Cape Mount County led to a gold rush; that year almost 31,000 ounces were exported, and nearly the same amount was exported in 1944. A decline in output subsequently occurred, but in 1950 exports still were above 12,000 ounces a year. Available data on gold for the 1950s and 1960s were based on purchases by the Bank of Monrovia, to which by law any gold mined in Liberia had to be sold. During these two decades the amount bought in most years was less than 2,000 ounces. Until the late 1970s purchases continued to remain small because the fixed purchase price was $35 an ounce at a time when open market prices were substantially higher. Gold mining was also restricted to Liberian citizens.

These regulations were altered by the Gold-Diamond Act of 1979, which revised the earlier 1958 legislation on diamond prospecting, mining, and trading to encompass gold as well. The law permitted foreigners to participate with Liberian owners of gold claims in developing the deposits. Approval was also given to brokers and dealers to purchase and export gold, and a gold appraisal office was established in the Ministry of Lands and Mines to facilitate exportation. Provision was also made to adjust the local price of gold regularly, depending on world prices. A thriving open market reportedly developed. From 1,086 ounces exported in 1979, the amount rose to 7,243 ounces in 1980 and to almost 19,200 ounces in 1981. A substantial drop occurred in 1982, but nearly 15,400 ounces were exported in 1983. The revised law had apparently resulted in some foreign investment, and one company was reported to have introduced mechanized digging equipment.

Diamonds

The existence of diamonds was reported in Liberia in the late nineteenth century, but these reports remain unconfirmed. The first confirmed discoveries were made in 1906, when some stones were recovered from alluvial deposits that were being panned for gold. Since then diamonds have been found in different parts of the country, but the major locations have been in Lofa and Nimba counties. Most mining was carried out on a small scale using crude equipment. Output remained quite small until after World War II. In 1950 finds in the lower parts of the Lofa River, as well as subsequent discoveries, resulted in mass diamond rushes that involved tens of thousands of potential prospectors. Many of the prospectors were workers from the rubber plantations, and their departure caused serious disruption in rubber collecting. In 1958 the government passed legislation to control prospecting, mining, and trading in diamonds. At the same time substantial fees were established for licenses.

Data on Liberia's diamond production have not been considered reliable. Liberia's use of the United States dollar as its unit of currency and domestic factors in neighboring Sierra Leone, where 3 substantial quantities of diamonds were also produced, were believed to result in extensive smuggling of diamonds into Liberia for sale. Both gem quality and industrial diamonds are found in Liberia, and annual export earnings vary depending not only on world price fluctuations but also on the relative quantities of each category of diamonds found. In 1970 some 800,000 carats having a value of $5.7 million were exported. In 1976 only 320,000 carats were exported, but earnings from fewer stones totaled $16.6 million, or 3.6 percent of the value of all exports from Liberia. Prices soared, and in 1979 the value of diamond exports reached a high of $39.6 million, or 7.4 percent of total export receipts. In 1983 some $17.2 million was received from the sale of 330,000 carats of diamonds. Under President Charles Taylor, Liberia was accused of supplying troops to support rebel forces in Sierra Leone's civil war. Taylor, a long-time ally of the Revolutionary United Front (RUF) in Sierra Leone, had supplied the rebels with arms in exchange for diamonds.

In 2000 the United Nations placed an 18-month ban on the international sale of the diamonds in an attempt to undermine the RUF, and in May of the following year it also imposed sanctions on Liberia. In 2003, because of the progress made in Liberia, President Gyude Bryant requested an end to the UN embargo on Liberian diamonds and timber, but the Security Council postponed such a move until the peace was more secure.

Iron

Liberia is rich in natural resources, especially in iron. Since 1951, Liberia was among the leading producers of iron ore in Africa, and Liberia is one of the principal exporters of iron ore in the world. Sizable reserves are found primarily in four areas: the Bomi Hills, the Bong Range, the Mano Hills, and Mount Nimba, where the largest deposits occur. Other minerals include diamonds, gold, lead, manganese, graphite, cyanite, and barite. There are also possible oil reserves off the coast.

The largest mining operation was the Liberian-American-Swedish Minerals Company (LAMCO), a joint venture that accounted for about half of Liberia's annual iron ore output. LAMCO's concession in the Nimba Range, near the border with Guinea, was given in 1953 but LAMCO only began shipping ore in 1963, when the port of Buchanan, which the company had constructed, finally opened for traffic. Their capacity was about 12 million tons of ore a year from a deposit at a proven reserve of some 250 million tons of 60 to 70 percent iron content. In the late 1970s ore output dropped to about 9 million tons, as American and European ore demands declined.

There also were proposals to move the potentially large output from ore deposits across the Guinean border via the LAMCO rail line to Buchanan if they were developed. In 1983 LAMCO's production had declined to 6.6 million tons. The second largest iron mining operation was in a 30-square mile concession located 50 miles north of Monrovia that had been granted in 1958 to the German-Liberian Mining Company. The firm was owned jointly in equal shares by the government and a consortium of German steel companies. Actual operations were carried out by the Bong Mining Company (BMC), and the ore was shipped to the German and Italian owned plants. The ore body had an average iron content of about 38 percent, which was increased to about 65 percent by concentration. Pelletizing, which required a high energy input, was also carried out. The profitability of the mine slumped as the 1970s progressed because of rising petroleum fuel costs. The production from the mine, which began in 1965, was shipped to the BMC pier at Monrovia port over a company-built rail line. By the early 1970s the output was over 5 million tons a year. From 1974 through 1975 output was generally over 6 million tons, and from the late 1970s to 1983 it averaged more than 7 million tons.

History

Declared a sovereign state in 1847, Liberia is unique among African countries. Next to Haiti, Liberia is the oldest black republic in the world and is the oldest republic in Africa. All the other countries in Africa have a history of colonization by white foreign nations. The colonial era of Liberia started when freed American slaves began to settle along the coast. The territory of Liberia was purchased from the native population for six firearms, one keg of gunpowder, three pairs of shoes, a box of beads and other trade goods.

The "Americo-Liberians," as these black settlers were known, never exceeded more than 5 percent of the country's total population. They settled in the urban centers they formed along the coast and maintained a society based on the cultural models they were familiar with back in the United States. The national majority of Liberia, the indigenous peoples, was eventually classified by the government into 16 different "tribes." Most of the native Africans were encouraged to remain in their homelands in the interior of the country; a region vaguely designated the "Hinterland." Exceptions were made, however, when inexpensive labor was needed on the large estates established by Americo-Liberians. Ironically, forced labor and other compulsory labor practices on these plantations were very often like the slavery experiences of the Americo-Liberians had left behind. The select minority of Americo-Liberians effectively excluded the indigenous majority from Liberia's social, political and economic life for over 130 years.

The Firestone Rubber Company began an experiment in rubber plantations in 1926, and the company's name would then become associated with Liberia for well over forty years. Prior to that, Liberia had a "century of survival," in that most of the efforts of the country were in preventing other countries and colonial powers from encroaching on their territory. After the strategic importance of Liberia was shown in WWII, America built a modern seaport and airport at the capital of Monrovia, named in honor of US President James Monroe.

In 1980, President Tolbert was assassinated in a coup led by Master Sergeant Samuel K. Doe. Twenty years of political violence, corruption and bad government ensued. A period of war and conflict lasted until 1997, and it is estimated that between 150,000 and 200,000 lives were lost, and many thousands of other people became refugees. Multiparty presidential and legislative elections held in July 1997 brought Charles Taylor to power. Under Taylor, the country remained economically depressed while he and his associates enriched themselves by looting their country's resources. In mid-2001, fighting erupted in Northern Liberia between anti-Taylor rebels and government forces. The civil war intensified during the next year, and the rebels continued to expand the war into other regions of Liberia in 2003. By mid-2003, the rebels controlled roughly two thirds of the country and were threatening to seize the capital, Monrovia.

In August, Charles Taylor resigned and went into exile, and he was succeeded for a short time by his vice president, Moses Blah. A peace agreement was signed with the two rebel groups, and several thousand West African peacekeepers, supported temporarily by an offshore U.S. force, arrived. In October 2003, the West African force was placed under UN command and was reinforced with troops from other nations; businessman Gyude Bryant became president of a new power-sharing government.

Despite the accord with the rebels, fighting initially continued in parts of the country; tensions among the factions in the national unity government also threatened the peace. Bryant's government was hindered by corruption and a lack of authority in much of Liberia, but the peace enabled the economy recover somewhat in 2004.

In the presidential election in the fall of 2004 former soccer star George Weah won the first round with 28% of the vote, but lost the runoff in November to Ellen Johnson Sirleaf, a politician and former World Bank official who received nearly 60% of the second round votes. Ellen Sirleaf thus became the first woman to be elected president of an African nation. At the same time a new national legislature was also elected, with no party securing a controlling position.

Civil war, corruption and government mismanagement have destroyed much of Liberia's economy, especially the infrastructure in and around Monrovia. Meanwhile, continued international sanctions on diamonds and timber exports limit growth prospects for the foreseeable future. Many businessmen, engineers and technicians have fled the country and the violence, taking their capital and their expertise with them. Some refugees have returned to Liberia, but many never will return.

Richly endowed with water, mineral resources, forests, and a climate favorable to agriculture, Liberia had been a producer and exporter of basic products- primarily raw timber and rubber. Local manufacturing, mainly foreign owned, had been small in scope. The government's changes have helped diffuse the political crisis, but have done little to encourage economic development. International donors, who are ready to assist reconstruction efforts, are withholding funding until Liberia's business and trading environment improves. A plan was created in October 2005 by the International Contact Group for Liberia to help ensure transparent revenue collection and allocation, and to put the brakes on government corruption. This was something that was lacking under the transitional government and this has also limited Liberia's economic recovery.

The reconstruction of infrastructure and the raising of incomes in this ravaged economy will largely depend on financial investment and technical assistance from other countries.

Abbreviations used:

(All links and URLs in this bibliography are current as of June 2006)

AGI: American Geological Institute, Alexandria, VA. See: www.agiweb.org

AGS: American Geographical Society Library, University of Wisconsin, Milwaukee Campus. See: http://www.amergeog.org

AMRS: Africa Mineral Resource Specialists Inc. 5 Blue Cedar, Suite 101, Littleton, Colorado 80127. AMRS owns a significant number of original geologic reports and maps for Liberia. This collection of data includes geologic reports, geologic maps and topographic maps. See: http://www.africaminerals.com/

DTIC: Defense Technical Information Center, Alexandria, VA. See: www.dtic.mil

FAO: Food and Agriculture Organization, United Nations, Rome, Italy. See: www.fao.org

GeoRef: American Geological Institute, Alexandria, VA. The American Geological Institute not only identified materials for the abstracting database, GeoRef, but also locates and supplies materials as a document delivery service. See: www.agiweb.org

ISBN: International Standard Book Number. This unique number can be used to identify and locate library holdings of a particular book or report title. See: http://www.isbn.org/standards/home/index.asp

ISSN: International Standard Serial Number. This unique number can be used to locate libraries which have subscriptions to this journal, magazine or serial. See: http://www.issn.org/

LC or LOC: Library of Congress, Geography and Map Dvision, Washington, DC. See: www.loc.gov

LCCN: Library of Congress Control Number. This is a unique number applied by the Library of Congress to identify individual publications. This number can be used to identify copies of this item in libraries held in the US and abroad. See: http://www.loc.gov/marc/lccn_structure.html

NTIS: National Technical Information Service, Alexandria, VA. See: www.ntis.gov

OCLC: Online Computer Library Center, Inc., Dublin, OH. See: www.oclc.org

UN: United Nations Library, New York, NY. See: www.un.org

USGS: US Geological Survey Library, Reston, VA www.usgs.gov

"An act approving the mining concession agreement entered into by and between the government of the Republic of Liberia and Liberian Gold and Diamond Corporation: approved October 12, 1976." Liberia. Ministry of Foreign Affairs. Monrovia: Ministry of Foreign Affairs. Description: 55 leaves; 28 cm. Subjects: Mining law-Liberia. Mineral industries-Law and legislation-Liberia. Notes: Cover title. OCLC: 54114366.

"L'Activité minière et activité industrièlle du Liberia telles qu'elles apparaissent á partir de rapport économique pour 1971." In: Industries et travaux d'outremer. Translated Title: "Industries and work of the overseas territories." Paris: R. Moreaux. Language: French. 19e année. No. 212, Jullet 1971. Pages 548-550. ISSN: 0019-9362; LCCN: 57-49914; OCLC: 2401146.

Adam, J.G. 1970. "The Vegetation of the Littoral Belt and Lagoons of Cape Palmas (Liberia)." (In French). Bulletin de la Société Botanique de France. Vol. 117, No 7/8, P 419-427. 1970. Illus. English Summary. Museum National D'histoire Naturelle, Paris (France). Abstract: Several vegetation zones are distinguished. No endemics were encountered. Descriptors: Cape; Lagoons; Liberia; Littoral Communities; Palmas; Vegetation; Zonation. ISSN: 0037-8941; OCLC: 1765763.

Aero Service Corporation. 1957. Planimetric map, Liberia. [Monrovia?]: Government of Liberia. 1 map in color; 51 x 51 cm. Descriptors: Liberia- Maps. Scale 1:1,000,000; Hotines rectified skew orthomorphic projection; (W 120--W 70/N 90--N 40). Notes: Relief shown by hachures. Alternate Liberia. "Published ... through the Joint Liberian-United States Commission for Economic Development; the United States Coast and Geodetic Survey; the Liberian Cartographic Service and the Bureau of Mines and Geology, Treasury Department, Republic of Liberia." "National boundary not shown. No field check." Compiled from shoran controlled mosaics of aerial photographs of 1953 by Aero Service Corporation, Philadelphia, PA. OCLC: 23531957; USGS Library.

Aero Service Corporation. 1957. Planimetric map, Liberia. Washington, D.C.: U.S. Coast and Geodetic Survey. 2 maps: col.; 82 x 58 cm. Contents: West half- East half. Scale 1:500,000; Hotines rectified skew orthomorphic proj.; (W 11030'--W 7030'/N 8030'--N 4030'). Notes: Relief shown by hachures. Alternate Liberia. "National boundary not shown. No field check." From shoran controlled mosaics of aerial photographs of 1953 by Aero Service Corporation, Philadelphia, PA. OCLC: 23531947.

"Africa." 1968. 1:2,000,000 Topographic Map Series, Monrovia Sheet, Sheet 16: US Army Map Service, 1968. AMRS, Inc. See: http://www.africaminerals.com/

Aisenstein, B. 1958. "Relations between geological site conditions and dam designs in west-Africa." International Geological Congress. 23rd Congress. Prague. Report. Abstracts. Pages 294-295.

Allen, G.M. 1930. Birds of Liberia. In: African Republic of Liberia. Massachusetts Department of Tropical Medicine and Institute of Tropical Medicine and Biology. Contributions. Volume 5, pages not cited.

Allen, G.M. and Coolidge, H. 1930. Mammals of Liberia. In: African Republic of Liberia. Massachusetts Department of Tropical Medicine and Institute of Tropical Medicine and Biology. Contributions. Volume 5, pages 569-622.

Allersma, Egge and Tilmans, Wiel M. K. 1993. "Coastal conditions in West Africa; a review." Ocean and Coastal Management. 19; 3, Pages 199-240. 1993. Elsevier Applied Science. Essex, United Kingdom. 1993. Descriptors: Africa; Atlantic Ocean; Benin; changes of level; coastal sedimentation; Ghana; Gulf of Guinea; human activity; Ivory Coast; Liberia; Niger Delta; Nigeria; North-Atlantic; review; sedimentation; shore features; shorelines; southeastern Liberia; southern Benin; southern Ghana; southern Nigeria; southern Togo; Togo; Volta Delta; West Africa; West Africa Coast; Engineering geology. References: 45; illustrations, including 9 tables, geol. sketch maps. Abstract: This paper describes the coastal system between Cape Palmas bordering the Ivory Coast and Liberia and Mount Cameroon, as a basis for regional coastal-erosion management. The area includes the rocky coasts of the Ivory Coast and Ghana, the long sandy beaches of Togo and Benin and the deltas of the Volta and Niger Rivers. Important elements of the system appear to be its geographical setting, its geological history, oceanological impacts, the sediment supply by rivers and coastal erosion, littoral transport and the rise of mean sea level. The morphological processes are influenced by human activities in the catchments of the rivers (dams), in the estuaries (dredging) and along the coasts (port construction). It appears that future human interventions and the impact of increased sea-level rise will require intergovernmental co-ordination in coastal-erosion management, as the various coasts are all part of one system. ISSN: 0964-5691.

American Colonization Society. 1800s. North west part of Montserrado County, Liberia: in ten mile squares. Liberia; Montserrado County. Year: 1800s. Description: 1 map on 5 sheets: manuscript, color; 89 x 62 cm. Series: American Colonization Society map collection; 21. Subjects: Montserrado County (Liberia)- Maps. Manuscript. Notes: Sheets taped together to form irregular shape. Pen-and-ink and watercolor. Available also through the Library of Congress Web site as a raster image. Scale [ca. 1:265,000]. LCCN: 96-684997; OCLC: 37986411. Access Path: http://hdl.loc.gov/loc.gmd/g8883m.lm000018

American Colonization Society. 1830s. "Map of Liberia, West Africa." G.F. Nesbitt & Company. N[ew] Y[ork]: Lith. G.F. Nesbitt & Co. Description: 1 map; 24 x 36 cm. Series: [American Colonization Society map collection; 1] Subjects: Tribes- Liberia- Maps. Liberia- Maps. Notes: Relief shown by hachures. Shows boundary of tribes, mission stations, colonist towns, and native towns. Available also through the Library of Congress Web site as a raster image. Map Info: Scale [ca. 1:1,380,000]. LCCN: 96-684983; OCLC: 37491634. Access Path: http://hdl.loc.gov/loc.gmd/g8881e.lm000007

American Geographical Society of New York and Herbert E. Budek Films and Slides (Firm). 1956. Liberia. New York: American Geographical Society; Hackensack, N.J.; Herbert E. Budek Co. Description: 1 filmstrip (50 fr.): color; 35 mm. and manual. Abstract: Outlines the principal elements of Liberia's economy; presents details of rubber production, picturing life on a rubber plantation; describes iron mining operations at Bomi Hills, and shows transport facilities. Contrasts views of the urban port of Monrovia with scenes of primitive native villages. Subjects: Liberia-

Economic conditions. LCCN: 57-2127; OCLC: 16952795.

"Analytic Image map research." 1983. USGS Professional Paper. Volume: 1375. Pages: 302-303. Report number: P1375. January 1, 1983. U. S. Geological Survey, Reston, VA, United States. Descriptors: aerial photography; Africa; Alaska; Arabian Peninsula; Asia; Bangladesh; Brazil; cartography; East Africa; Ethiopia; geophysical methods; Grand Canyon; imagery; Indian Peninsula; Landsat; Liberia; maps; Mexico; Nepal; Pakistan; radar methods; remote sensing; research; Saudi Arabia; South America; United States; USGS; West Africa; Yemen: Applied geophysics. ISSN: 1044-9612.

Anderson, Benjamin. 1879. "Map of the Republic of Liberia: Constructed from Authentic Charts & Original Surveys." Monrovia. Scale 1:1,013,760. Verified October 3, 1879 [by] G.W. Gibson, Secretary of State. Library of Congress, Geography & Map Division.

Anonymous. No date given. "Map Showing Major Traverse Routes, Distribution of Kimberlitic and Diamond Indicator Minerals, Extent and Intensity of Alluvial Diamond Activities in the Southwestern Section of the Komgba Forest, Lofa County, Liberia." No place of publication given. AMRS, Inc. See: http://www.africaminerals.com/

Appiah, Henry. 1990. "Gold mineralization in the Prestea-Obuasi Belt." In: 15th colloquium of African geology--15 (super e) colloque de geologie africaine. Publication Occasionnelle - Centre International Pour la Formation et les Echanges Geologiques = Occasional Publication - International Center for Training and Exchanges in the Geosciences. 20; Pages 332. 1990. Centre International pour la Formation et les Echanges Geologiques (CIFEG). Paris, France. 1990. Conference: 15th colloquium of African geology--15 (super e) colloque de geologie africaine. Nancy, France. Sept. 10-13, 1990. Descriptors: Africa; gold ores; Liberia; Liberian Shield; metal ores; mineral deposits, genesis; Prestea Obuasi Belt; structural controls; West Africa; Economic geology: geology of ore deposits. ISSN: 0769-0541.

Archambault, Jean. 1961. Les Eaux Souterraines de l'Afrique Occidentalte. Translated title: "Subterranean water of West Africa." Nancy: Berger-Levrault. Paris? Bureau central d'ètudes pour les èquipements d'outre-mèr. 137 pages, illustration, maps (2 fold. in pocket) 28 cm. OCLC: 17944792.

"Arms for diamonds.(Liberia and Burkina Faso accused of aiding rebels in Sierra Leone though trade in arms for diamonds)." The Economist (UK). August 5, 2000: page 6(1). Subjects: Liberia. Diamond mining. Diamond industry- economics.

Armstrong, Robert P. 1962. "The Role of Foreign Concessions in the Economy of Liberia." Northeastern University Economic Survey of Liberia. Northeastern University, Evanston, Illinois. 231 pages. OCLC: 260380.

Aseyeva, Ye. A.; Furtes, V. V. 1998. "Sravnitel'naya kharakteristika verkhneproterozoyskikh otlozheniy okrain liberiyskogo i ukrainskogo shchitov po mikrofossiliyam." Translated title: "Correlation of the upper Proterozoic deposits of Liberian and Ukrainian shields from microfossil data." Geologichnyy Zhurnal (1995) = Geological Journal. 1998; 3-4, Pages 81-86. 1998. Natsional'na Akademiya Nauk Ukrayini, Institut Geologichnikh Nauk. Kiev, Ukraine. 1998. Language: Russian; Summary Language: English; Ukrainian. Abstract: The first find of fossil microorganisms from the Mali-suite deposits of the northwestern outskirt of Liberian shield are analogous to ones from the Proterozoic deposits (vendian) of the Ukrainian shield. This analogy allows us to carry out the transcontinental correlation of these open-casts. Descriptors: Africa; algae; argillite; biostratigraphy; clastic rocks; Commonwealth of Independent States; correlation; Europe; Liberia; Liberian Shield; Mali Suite; microfossils; Neoproterozoic; Plantae; Precambrian; Proterozoic; Russian Platform; sandstone; sedimentary cover; sedimentary rocks; Ukraine; Ukrainian Shield; upper Precambrian; Vendian; West Africa; Stratigraphy. References: 9; illustrations. ISSN: 1025-6814.

Ashmun, Jehudi. 1794-1828. 1830. "Map of the west coast of Africa from Sierra Leone to Cape Palmas, including the Colony of Liberia." American Society for Colonizing the Free People of Colour of the United States. Annual report. Philadelphia: A. Finley. Description: 1 map. Subjects: Liberia- Maps. Monrovia (Liberia)- Maps. Africa, West- Maps. Map Info: 1:2,534,400; (W 13020'- -W 7020'/N 8030'--N 4015'). Notes: Fold-out plate detached from: The thirteenth annual report of American Society for Colonizing the Free People of Colour of the United States.- Washington, 1830. Includes descriptive and historical "Remarks." Heights shown pictorially; depths shown with gradient tints of gray. Inset: Plan of the town of Monrovia. Compiled chiefly from the surveys and observations of the late Rev. J. Ashmun; J. H. Young. OCLC: 11083752.

Assessment of Energy Options for Liberia. Final Report. 1983. Oak Ridge National Lab., Oak Ridge, TN, United States. Funded by: Department of Energy, Washington, DC. Nov 83. 101p. Contract number: W7405ENG26; Report number: ORNL5989. Descriptors: *Liberia; Cogeneration; Combustion; Commercial Sector; Data Compilation; Energy Analysis; Energy Demand; Energy Source Development; Energy Supplies; Evaluation; Forecasting; Gasification; Hydroelectric Power; Industry; Petroleum Refineries; Planning; Residential Sector; Theoretical Data; Transportation Sector; Wood; Wood Fuels; Wood Fuel Power Plants; Energy conversion non propulsive Conversion techniques; Energy Miscellaneous energy conversion and storage; Energy Policies regulations and studies; Energy use supply and demand. Abstract: Under funding from the U.S. Agency for International Development (USAID), the Oak Ridge National Laboratory provided energy planning assistance to the National Energy Committee of the Government of Liberia (GOL), West Africa, during a period of one year ending March 31, 1983. This report outlines the scope of activities of the joint GOL/USAID project and summarizes the major findings by Liberian and U.S. project participants. The study included an examination of current energy use by sector and fuel type, projections of future energy demands, and a preliminary evaluation of a variety of alternative energy resource and technology options for Liberia. The primary finding is that Liberia has significant opportunities for the substitution of indigenous energy resources for imported petroleum. The principal candidates are wood energy and hydroelectric power. The major alternatives for wood are gasification technology for small-scale electric and non-electric applications (e.g., those under about 25-gigajoule/hour input requirements) and wood-fired steam electric generation for larger scale applications where hydroelectric power is unattractive. For major hydroelectric development the principal candidates are the St. Paul River Proposal and the Mano River Proposal. The Mano River Proposal is the smaller of the two and would meet Monrovia area electrical grid requirements and some iron ore mine demand for about the next two decades. An additional important finding of this study is that the Monrovia Petroleum refinery is highly uneconomical and should be closed and petroleum products imported directly. 25 tables. NTIS Number: DE84004110XSP.

Avery family. 1916-1976. Johnston and Virginia Hall Avery private papers. Archival Material ca. 3376 items. Descriptors: Governors -- North Carolina. Senators -- United States -- Correspondence. Natural resources -- Africa -- Liberia. Abstract: The collection consists of carbon copies of correspondence and papers which reflect Johnston Avery's involvement with several important issues of his time. The majority of papers consist of correspondence written or received by Avery including holograph, typewritten and typewritten carbon copy letters. Important correspondents include William V.S. Tubman, President of Liberia from 1944 to 1971 and other Liberian government officials; Robert R. Reynolds, U.S. Senator from North Carolina from 1933 to 1945; J.C.B. Ehringhaus, North Carolina governor from 1933 to 1937; and various businessmen involved in the LAMCO Joint Venture. A few items were not addressed to Mr. Avery. These items include letters written by or addressed to his wife, Virginia Hall Avery. Notes: Bio/History: Mr. Johnston Avery, a North Carolina native, pursued several careers. He was a newspaperman, political campaign aide, government official and businessman. He assisted Senator Robert R. Reynolds in his campaign for repeal of the Eighteenth Amendment (Liquor Prohibition Amendment). He served as Assistant Chief of Decartelization Branch of the U.S. Office of Military Government (OMGUS) in Berlin following World War II. He became convinced that OMGUS was blocking attempts by the Decartelization Branch to carry out its mission of breaking up the German cartels, he resigned and brought charges against OMGUS. A Congressional investigation ensued which substantiated these charges. Mr. Avery became involved in the economic development and exploitation of natural resources of Africa, specifically in Liberia. With the State Department, he was Assistant Administrator of the Point Four Program, which was established to aid developing countries acquire technical expertise. He resigned this position to become President of the Liberian American Swedish Minerals Company (LAMCO) Joint Venture, a position he held until his death. OCLC: 49886402.

Axelrod, J. M.; Carron, M. K.; and Thayer, Thomas P. 1952. "Phosphate mineralization at Bomi Hill and Bambuta, Liberia, west Africa." American Mineralogist. 37; 11-12, Pages 883-909. 1952. Mineralogical Society of America. Washington, DC, United States. Abstract: Iron phosphate minerals which cement talus ore below cliffs formed by massive magnetite- hematite deposits at Bomi hill, Liberia, and also occur in place in fissures and caves in the ore both at Bomi hill and Bambuta, were formed by the interaction of bat dung and iron oxides. The minerals include leucophosphite (previously known only from Western Australia), phosphosiderite, and strengite. Analyses of the leucophosphite are included. Descriptors: Africa; Bomi hills; Bambuta area; Liberia; mineral data; Phosphate minerals; West Africa. Illustrations. No. 3-4, pages: 284. ISSN: 0003-004X.

Axelrod, J. M; Carron, M. K; Milton, C.; and Thayer, T. P. 1951. "Phosphate mineralization at Bomi hills and Bambuta, Liberia, west Africa." Geological Society of America Bulletin. 62; 12, Part 2, Pages 1421-1422. 1951. Geological Society of America (GSA). Boulder, CO, United States. Descriptors: Africa; Bomi hills, Bambuta area; Liberia; mineral data; Phosphate minerals; West Africa. ISSN: 0016-7606.

Ballon, H. J. 1981. "Die Bohr- und Sprengarbeit bei der Bong Mining Company in Liberia Westafrika; II." Translated title: "Drilling and explosive work at the Bong Mining Company in Liberia, West Africa; II." Erzmetall. 34; 3, Pages 147-152. 1981. Dr. Riederer-Verlag. Stuttgart, Federal Republic of Germany. Language: German; Summary Language: French; English. Descriptors: Africa; Bong Mining Co; companies; economic geology; economics; environmental geology; exploitation; explosions; geologic hazards; iron ores; Liberia; metal ores; mining; mining geology; open pit mining; surface mining; West Africa. References: 5; illustrations, including 2 tables, block diag. ISSN: 0044-2658.

Balsley, J. R. (investigator). 1979. Liberia. U. S. Geological Survey Professional Paper #P1150. Pages 326. 1979. U. S. Geological Survey. Reston, VA. Descriptors: Africa; areal geology; Liberia; maps; West Africa; Geologic maps. ISSN: 1044-9612.

Bardet, M. G. 1974. "Geologie du diamant; deuxieme partie; Gisements de diamant d'Afrique." Translated title: "The geology of diamonds; 2nd part, Diamond deposits of Africa." Memoires du B.R.G.M. 83; 2, 1974. Bureau de Recherches Geologiques et Minieres, (BRGM). Paris, France. Pages: 229. Language: French. Abstract: Contains a separate printout on new ideas and data on kimberlites and diamonds. Descriptors: Africa; Angola; Botswana; Cameroon; Central Africa; Central African Republic; Congo Democratic Republic; diamonds; distribution; East Africa; economic geology; exploration; Gabon; genesis; Ghana; Guinea; igneous rocks; Ivory Coast; kimberlite; Lesotho; Liberia; Mali; maps; Namibia; plutonic rocks; production; regional; resources; Sahara; Sierra Leone; South Africa; Southern Africa; Tanzania; tectonic; tectonics; ultramafics; West Africa; Zaire; Zimbabwe. ISSN: 0071-8246.

Barnes R. F. W. and Dunn, A. 2002. "Estimating forest elephant density in Sapo National Park (Liberia) with a rainfall model." African Journal of Ecology 40, no. 2 (2002) p. 159-163. Abstract: The number of elephant dung-piles lying on the forest floor is a function of the number of elephants present and the rainfall in the 2 preceding months. We present the results of a stochastic model that describes this relationship and we show how it can be used to estimate elephant numbers. The data from a survey in Sapo NP (Liberia) in 1989 are used as an example. The dung-pile density was estimated at 152 km-2 with confidence interval from 72 to 322, and the number of elephants was estimated to be 313 with confidence interval from 172 to 617. References: 14. Descriptors: Animal ecology; mammals; population ecology; population density. ISSN: 0141-6707.

Barron, W. F.; Hobbs, B. F.; Samuels, G.; and Kawah, L. M. 1985. Background Paper on Electrical Services Provided by the Liberia Electricity Corporation (LEC). Performer: Oak Ridge National Lab., Oak Ridge, TN, United States. Funded by: Liberia Electricity Corp., Monrovia. Planning Dept. Funded by: Department of Energy, Washington, DC. Contract number: AC0584OR21400. July 1985. 43p. Report number: ORNLTM9425. Advisory: Portions of this document are illegible in microfiche products. Original copy available until stock is exhausted. Descriptors: Data Compilation; Electric Power; Energy Consumption; Fuel Consumption; Power Demand; Power Generation; Prices; Sales; Electric Utilities; Liberia. Abstract: This report is one of a series of project papers providing background information for an assessment of energy options for Liberia, West Africa; it presents data on electrical services in Liberia (as of early 1983) with primary emphasis on the operations of the Liberia Electricity Corporation (LEC). The LEC is a semiautonomous agency owned by the Government of Liberia that has primary responsibility for generating electricity throughout Liberia. The LEC system consists of a central grid covering an area roughly 175 by 100 km with Monrovia as its focal point, and nine rural stations serving ten towns. The central grid has a total capacity of 177 MW (64 hydro and 113 diesel engines and gas turbines) and produced 378 million kWh in 1981. The rural stations with a total capacity of 13 MW (all diesels) produced 27 million kWh in 1981. Information provided by this paper includes historical sales data by customer class, growth in demand, hourly load data, petroleum consumption, prices, and problems. Major problems include uncollected bills, illegal hookups, inoperable generating equipment, and fuel shortages. (ERA citation 10:048914). NTIS: DE85018123XSP.

"Barytes in Liberia." 1968. Mining Journal. Number 270, page 423. ISSN: 0026-5225; OCLC: 02438984.

Bassler, Fritz. 1958. Unterschlossene Wasserkräfte in Liberia. Translated title: "Under closed water power in Liberia" Stuttgart: Die Wasserwirtschaft. Volume 46, no. 6, pages 141-150.

Basso, Zh. (Bassot, J) and Gottin, G. (Hottin, G). 1983. "Orudeneniye, svyazannoye s dokembriyskimi granitoidnymi porodami Zapadnoy Afriki." Translated title: "Mineralization of Precambrian granites in West Africa." In: Metallogeniya dokembriyskikh granitoidov. Translated title: Metallogeny of Precambrian granites. Luchitskiy, I.V. (editor). Pages 149-195. 1983. Nauka. Moscow, Russian Federation. 1983. Language: Russian; Summary Language: French. Title on verso title page: Metallogeny of Precambrian granitoids. At head of title: Akademiia nauk SSSR. Institut litosfery. Mezhdunarodnaia programma "Litosfera." Summary in English; table of contents also in English. Includes bibliographies. Descriptors: Africa; Archean; Benin; beryl; Burkina Faso; chain silicates; clinopyroxene; ferruginous quartzite; granites; Guinea; igneous rocks; Ivory Coast; lead zinc deposits; Liberia; Mauritania; metal ores; metamorphic rocks; mineralization; molybdenum ores; Niger; Nigeria; pegmatite; plutonic rocks; polymetallic ores; Precambrian; pyroxene group; quartzite; rare earth deposits; ring silicates; Senegal; Sierra Leone; silicates; spodumene; West Africa. References: 28; illustrations, including geol. sketch maps. LCCN: 84-188545; OCLC: 12052985.

Battelle Memorial Institute; Columbus Laboratories; and United States. Agency for International Development. 1970. Summary report on transport improvements and Governmental fiscal/administrative policies in relation to the development of timber and other resources in southeast Liberia, to the United States Agency for International Development. Columbus: Battelle Memorial Institute, Columbus Laboratories. 544 pages in various pagings: illustration; 29 cm. Research report- Battelle Memorial Institute, Columbus Laboratories; Variation: Battelle Memorial Institute; Columbus Laboratories. Research report. LCCN: 77-366146; OCLC: 3445525.

Bauseler, K. H. 1970. "Planning and construction of a blending plant for crude ore and its influence on mining operations." Bulletin of the Geological, Mining and Metallurgical Society of Liberia. Volume 4; Pages 56-70. 1970. Notes: Vol. 4. Geological, Mining and Metallurgical Society. Monrovia, Liberia. Descriptors: Africa; Bong mine; economic geology; Liberia; mineral resources; production; West Africa; Economic geology, general. Illustrations. ISSN: 0367-4819.

Behrendt, John Charles. 2005. Interview. Editor: Shoemaker, Brian. March 20, 2006, July 15, 2005, March 14, 2000. Note: John Behrendt first visited Antarctica in November, 1956, where he spent the winter at Ellsworth Station as an assistant seismologist to Edward Thiel, and he remained there until January, 1958. He became interested in geophysics while a student at the University of Wisconsin in Madison. Some earlier research at Lemon Creek Glacier in the Juneau Ice Field at Juneau, Alaska, had given him valuable experience with crevasses. Behrendt spent some more time in Antarctica in several expeditions, and he returned home in February, 1962. He mentioned for the first time that he had been married the day before he left Madison to start the traverse. He left the University of Wisconsin and went to work for the U.S. Geological Survey in 1964. He was also a new father by this time. He thought he was through with Antarctica, but found himself there once again for the 1965-66 season. He led the geophysical part of a major field camp operation. The party went to the Pensacola Mountains, had a camp in the Neptune Range, and did aeromagnetic surveys of the entire Pensacola Mountains, including the Dufek Massif. Much gravity work was done. They gathered a large amount of geologic and topographic mapping data, and compiled these into maps. His duties were completed in a month, and he returned home to work for the USGS. Later he spent several years working in Liberia, accompanied by his wife, and by then, two young sons. Later he moved to Wood's Hole, Massachusetts, where he became Chief of the Atlantic Gulf Branch of Marine Geology. He made several brief trips to Antarctica between 1978 and 1995. Altogether he made 12 trips to Antarctica. Behrendt joined the U.S. delegation that helped draft the Convention on the Regulation of Antarctic Mineral Resource Activities (CRAMRA), which, as later amended, was partially blended into the Environmental Protocol to the Antarctic Treaty. The following season, 1984-85, he worked with the German-Antarctic North Victoria Land Expedition and flew aeromagnetic surveys over northern Victoria Land. Various countries have participated in Antarctic research in recent years, including Brazil, Japan, Germany, USSR, Russia, and France. Italy has been especially active. But the U.S. Geological Survey has cut its Antarctic and Arctic research in the last 5 years [prior to 2000] or so. His last trip to Antarctica was in the 1994-95 season. In recent years there has been environmental concerns expressed in some quarters about the effect of tourism and fishing in or near Antarctica. Behrendt feels that by and large the tour operators have been responsible, but some fishing ships have been pirating the protected Antarctic bonefish. Chilean sea bass sold in the US is probably pirated from Antarctic waters. Behrendt comments on several of the cruise ships visiting Antarctica, including the Polar Queen and the Polar Duke. Since 1996 he has been a Fellow at the University of Colorado at the Institute of Arctic and Alpine Research, and has continued his researches on Antarctica to the present day [March, 2000]. He has served as a leading advisor to the State Department on the Environmental Protocol to the Antarctic Treaty, and also worked on CRAMRA, the convention on the Regulation of Antarctic Mineral Resource activities. He attended 22 international meetings between 1977 and 1995. He reports that the U.S. continues to operate the South Pole Station, and significant scientific research continues from various groups and countries. The icebreaker, the Nathaniel Palmer, continues to do very good research. NOAA currently is not very active in Antarctica. Behrendt published a book in 1957, "Innocents on the Ice: A Memoir of Antarctic Exploration" and is working on another one. He hopes to return to Antarctica again. Notes: Knowledge Bank at Ohio State University. Funded by a grant from the National Science Foundation. 3 audio tapes available in the OSU Archives. Polar Oral History Program. Record Group Number: 56.36. Subjects: Antarctica- Discovery and exploration- Interviews; Behrendt, John C., 1932-Interviews. URL: http://hdl.handle.net/1811/6057.

Behrendt, John Charles. 1970. "An aeromagnetic and aeroradioactivity survey of Liberia." Geophysics. 35; 6, Pages 1162. 1970. Society of Exploration Geophysicists. Tulsa, OK, United States. Descriptors: Africa; geophysical methods; geophysical surveys; Liberia; magnetic; magnetic methods; radioactivity; radioactivity methods; regional; surveys; West Africa; Applied geophysics. ISSN: 0016-8033.

Behrendt, John Charles. 1970. "An aeromagnetic and aeroradioactivity survey of Liberia." In: Searching the seventies, from Moho to Mars. Pages 100-101. 1970. Society of Exploration Geophysicists. New Orleans, LA, United States. Conference: Society of Exploration Geophysicists, 40th Annual International Meeting. New Orleans, LA, United States. Nov. 8-12, 1970. Descriptors: Africa; geophysical methods; geophysical surveys; Liberia; magnetic methods; radioactivity; radioactivity methods; regional; surveys; West Africa; Applied geophysics.

Behrendt, John Charles and Wotorson, Cletus S.[1] 1976. "Aeromagnetic map of the Zorzor quadrangle, Liberia." Washington: U.S. Geological Survey. 7 leaves: maps (2 folded); 27 cm. Open-file report (Geological Survey (U.S.)); 1601. Notes: Project report, Liberian investigations (IR) LI-69 C. Referred to in press release dated August 2, 1971. Bibliography: leaves 6-7. USGS Library.

Behrendt, John Charles and Wotorson, Cletus S. 1974. "Aeromagnetic map of the Bopolu Quadrangle, Liberia." US Geological Survey, Reston, VA, United States. Miscellaneous Investigations Series Map. January 1, 1974. Report number: I-0772-B. Map Type: magnetic survey map. Annotation: Lat 7 degrees to 8 degrees, long 10 degrees to 11 degrees. Scale 1:250,000 (1 inch = about 4 miles). Sheet 28 1/2 by 35 1/2 inches. Descriptors: Africa; airborne; Bopolu Quadrangle; geophysical methods; geophysical surveys; Liberia; magnetic methods; maps; northwest; surveys; USGS; West Africa. ISSN: 0160-0753.

Behrendt, John Charles and Wotorson, Cletus S. 1974. "Aeromagnetic map of the Buchanan Quadrangle, Liberia." U. S. Geological Survey, Reston, VA, United States. Miscellaneous Investigations Series Map. January 1, 1974. Report number: I-0778-B. Map Type: magnetic survey map. Annotation: Lat 5 degrees to 6 degrees, long 9 degrees to 10 degrees. Scale 1:250,000 (1 inch = about 4 miles). Sheet 29 by 34 1/2 inches. Descriptors: Africa; airborne; Buchanan Quadrangle; geophysical methods; geophysical surveys; Liberia; magnetic methods; maps; southwest; surveys; USGS; West Africa. ISSN: 0160-0753.

Behrendt, John Charles and Wotorson, Cletus S. 1974. "Aeromagnetic map of the Gbanka Quadrangle, Liberia." U. S. Geological Survey, Reston, VA, United States. Miscellaneous Investigations Series Map. January 1, 1974. Report number: I-0776-B. Map Type: magnetic survey map. Annotation: Lat 6 degrees to 7 degrees, long 9 degrees to 10 degrees. Scale 1:250,000 (1 inch = about 4 miles). Sheet 26 by 29 inches. Descriptors: Africa; airborne; central; Gbanka Quadrangle; geophysical methods; geophysical surveys; Liberia; magnetic methods; maps; surveys; USGS; West Africa. ISSN: 0160-0753.

---

[1] Cletus Segbe Wotorson (born 13 March 1937) is a Liberian politician and geologist. Director of the Liberian Geological Survey, 1973-1975.

Behrendt, John Charles and Wotorson, Cletus S. 1974. "Aeromagnetic map of the Harper Quadrangle, Liberia." U. S. Geological Survey, Reston, VA, United States. Miscellaneous Investigations Series Map. January 1, 1974. Report number: I-0780-B. Map Type: magnetic survey map. Scale 1:250,000 (1 inch = about 4 miles). Sheet 29 by 41 inches. Descriptors: Africa; airborne; geophysical methods; geophysical surveys; Harper Quadrangle; Liberia; magnetic methods; maps; southeast; surveys; USGS; West Africa. ISSN: 0160-0753.

Behrendt, John Charles and Wotorson, Cletus S. 1974. "Aeromagnetic map of the Juazohn Quadrangle, Liberia." U. S. Geological Survey, Reston, VA, United States. Miscellaneous Investigations Series Map. January 1, 1974. Report number: I-0779-B. Map Type: magnetic survey map. Scale 1:250,000 (1 inch = about 4 miles). Sheet 29 by 40 1/2 inches. Descriptors: Africa; airborne; geophysical methods; geophysical surveys; Juazohn Quadrangle; Liberia; magnetic methods; maps; southeast; surveys; USGS; West Africa. ISSN: 0160-0753.

Behrendt, John Charles, and Wotorson, Cletus S. 1974. "Aeromagnetic map of the Monrovia Quadrangle, Liberia." U. S. Geological Survey, Reston, VA, United States. Miscellaneous Investigations Series Map. January 1, 1974. Report number: I-0775-B. Map Type: magnetic survey map. Scale 1:250,000 (1 inch = about 4 miles). Sheet 29 by 36 inches. Descriptors: Africa; airborne; geophysical methods; geophysical surveys; Liberia; magnetic methods; maps; Monrovia Quadrangle; surveys; USGS; west; West Africa. ISSN: 0160-0753.

Behrendt, John Charles and Wotorson, Cletus S. 1974. "Aeromagnetic map of the Sanokole Quadrangle, Liberia." U. S. Geological Survey, Reston, VA, United States. Miscellaneous Investigations Series Map. January 1, 1974. Report number: I-0774-B. Map Type: magnetic survey map. Annotation: Lat 7 degrees to 8 degrees, long 8 degrees to 9 degrees. Scale 1:250,000 (1 inch = about 4 miles). Sheet 26 by 29 inches. Descriptors: Africa; airborne; geophysical methods; geophysical surveys; Liberia; magnetic methods; maps; northeast; Sanokole Quadrangle; surveys; USGS; West Africa. ISSN: 0160-0753.

Behrendt, John Charles and Wotorson, Cletus S. 1974. "Aeromagnetic map of the Voinjama Quadrangle, Liberia." U. S. Geological Survey, Reston, VA, United States. Miscellaneous Investigations Series Map. January 1, 1974. Report number: I-0771-B. Map Type: magnetic survey map. Annotation: Scale 1:250,000 (1 inch = about 4 miles). Sheet 29 by 32 inches. Descriptors: Africa; airborne; geophysical methods; geophysical surveys; Liberia; magnetic methods; maps; north; surveys; USGS; Voinjama Quadrangle; West Africa. ISSN: 0160-0753.

Behrendt, John Charles and Wotorson, Cletus S. 1974. "Aeromagnetic map of the Zorzor Quadrangle, Liberia." US Geological Survey, Reston, VA, United States. Miscellaneous Investigations Series Map. January 1, 1974. Report number: I-0773-B. Map Type: magnetic survey map. Annotation: Lat 7 degrees to 8 degrees, long 9 degrees to 10 degrees. Scale 1:250,000 (1 inch = about 4 miles). Sheet 26 by 29 inches. Descriptors: Africa; airborne; geophysical methods; geophysical surveys; Liberia; magnetic methods; maps; north; surveys; USGS; West Africa; Zorzor Quadrangle. ISSN: 0160-0753.

Behrendt, John Charles and Wotorson, Cletus S. 1974. "Geophysical surveys of Liberia with tectonic and geologic interpretations." U. S. Geological Survey, Reston, VA, United States. Pages:

33. Report number: Professional Paper 810. January 1, 1974. Map Scale: 1:1,000,000. Map Type: tectonic map, geophysical survey. Maps. Annotation: Plates in pocket. Illustrations. Notes: Part of illustrative matter in pocket. Bibliography: p. 32-33. Total magnetic intensity map of Liberia is plate #2. 1 sheet, black and white. 1:1,000,000. Descriptors: Africa; airborne; Bouguer anomalies; dikes; evolution; faults; geophysical methods; geophysical surveys; gravity anomalies; gravity methods; intrusions; Liberia; magnetic anomalies; magnetic methods; maps; Phanerozoic; Precambrian; radioactivity methods; regional; structure; surveys; tectonic; tectonics; USGS; West Africa. ISSN: 1044-9612; OCLC: 1365990.

Behrendt, John Charles and Wotorson, Cletus S. 1974. "Magnetic and radiometric data of Liberia, West Africa, area block 1." U.S. Geological Survey, Reston, VA, United States. January 1, 1973. Date (other): January 1, 1974. Annotation: Magnetic tape. Descriptors: Africa; geophysical surveys; Liberia; surveys; USGS; West Africa. NTIS Report number: PB-223 208/0.

Behrendt, John Charles and Wotorson, Cletus S. 1974. "Magnetic and radiometric data of Liberia, West Africa, area block 2." US Geological Survey, Reston, VA, United States. Availability: NTIS, Springfield, VA, United States. January 1, 1973. Date (other): January 1, 1974. Annotation: Magnetic tape. Descriptors: Africa; geophysical surveys; Liberia; surveys; USGS; West Africa. NTIS Report number: PB-223 209/8.

Behrendt, John Charles and Wotorson, Cletus S. 1974. "Magnetic and radiometric data of Liberia, West Africa, area block 3." U.S. Geological Survey, Reston, VA, United States
Availability: NTIS, Springfield, VA. January 1, 1973. Date (other): January 1, 1974. Annotation: Magnetic tape. Descriptors: Africa; geophysical surveys; Liberia; surveys; USGS; West Africa. NTIS Report number: PB-223 210/6

Behrendt, John Charles and Wotorson, Cletus S. 1974. "Magnetic and radiometric data of Liberia, West Africa, area block 4." US Geological Survey, Reston, VA, United States. January 1, 1973. Date (other): January 1, 1974. Annotation: Magnetic tape. Descriptors: Africa; geophysical surveys; Liberia; surveys; USGS; West Africa. NTIS Report number: PB-223 211/4.

Behrendt, John Charles and Wotorson, Cletus S. 1974. "Magnetic and radiometric data of Liberia, West Africa, area block 5." US Geological Survey, Reston, VA, United States. January 1, 1973. Date (other): January 1, 1974. Annotation: Magnetic tape. Descriptors: Africa; geophysical surveys; Liberia; surveys; USGS; West Africa. NTIS Report number: PB-223 212/2.

Behrendt, John Charles and Wotorson, Cletus S. 1974. "Magnetic and radiometric data of Liberia, West Africa, area block 6." US Geological Survey, Reston, VA, United States. January 1, 1973. Date (other): January 1, 1974 .Annotation: Magnetic tape. Descriptors: Africa; geophysical surveys; Liberia; surveys; USGS; West Africa. NTIS Report number: PB-223 213/0.

Behrendt, John Charles and Wotorson, Cletus S. 1974. "Magnetic and radiometric data of Liberia, West Africa, area block 7." US Geological Survey, Reston, VA, United States. January 1, 1973. Date (other): January 1, 1974. Annotation: Magnetic tape. Descriptors: Africa; geophysical surveys; Liberia; surveys; USGS; West Africa. Serial Report Available Only through NTIS.

Behrendt, John Charles and Wotorson, Cletus S. 1974. "Magnetic and radiometric data of Liberia, West Africa, area block 8." US Geological Survey, Reston, VA, United States. January 1, 1973. Date (other): January 1, 1974. Annotation: Magnetic tape. Descriptors: Africa; geophysical surveys; Liberia; surveys; USGS; West Africa. NTIS Report number: PB-223 215/5.

Behrendt, John Charles and Wotorson, Cletus S. 1974. "Magnetic and radiometric data of Liberia, West Africa, area block 9." US Geological Survey, Reston, VA, United States. Availability: NTIS, Springfield, VA, United States. January 1, 1973. Date (other): January 1, 1974. Annotation: Magnetic tape. Descriptors: Africa; geophysical surveys; Liberia; surveys; USGS; West Africa. NTIS Report number: PB-223 216/3.

Behrendt, John Charles and Wotorson, Cletus S. 1974. Magnetic and radiometric data of Liberia, West Africa, area block 10. U.S. Geological Survey, Reston, VA, United States. January 1, 1973. Date (other): January 1, 1974. Annotation: Magnetic tape. Descriptors: Africa; geophysical surveys; Liberia; surveys; USGS; West Africa. NTIS Report number: PB-223 217/1

Behrendt, John Charles and Wotorson, Cletus S. 1974. "Simple Bouguer gravity map of the Monrovia Quadrangle, Liberia." U. S. Geological Survey, Reston, VA, United States. Miscellaneous Investigations Series Map. January 1, 1974. Report number: I-0775-E. Map Scale: 1:250,000. Map Type: gravity survey map. Annotation: Scale 1:250,000 (1 inch = about 4 miles). Sheet 27 by 33 inches. Descriptors: Africa; Bouguer anomalies; geophysical methods; geophysical surveys; gravity anomalies; gravity methods; Liberia; maps; Monrovia Quadrangle; surveys; USGS; West Africa. ISSN: 0160-0753.

Behrendt, John Charles and Wotorson, Cletus S. 1974. "Total-count gamma radiation map of the Bopolu Quadrangle, Liberia." U. S. Geological Survey, Reston, VA, United States. Miscellaneous Investigations Series Map. January 1, 1974. Report number: I-0772-C. Map Type: geophysical survey. Map. Annotation: Lat 7 degrees to 8 degrees, long 10 degrees to 11. Scale 1:250,000 (1 inch = about 4 miles). Sheet 29 by 35 inches. Descriptors: Africa; airborne; Bopolu Quadrangle; gamma ray; geophysical methods; geophysical surveys; Liberia; maps; northwest; radioactivity methods; surveys; USGS; West Africa. ISSN: 0160-0753.

Behrendt, John Charles and Wotorson, Cletus S. 1974. "Total-count gamma radiation map of the Buchanan Quadrangle, Liberia." US Geological Survey, Reston, VA, United States. Miscellaneous Investigations Series Map. January 1, 1974. Report number: I-0778-C. Map Type: geophysical survey map. Scale 1:250,000 (1 inch = about 4 miles). Sheet 29 by 31 inches. Descriptors: Africa; airborne; Buchanan Quadrangle; gamma ray; geophysical methods; geophysical surveys; Liberia; maps; radioactivity methods; southwest; surveys; USGS; West Africa. ISSN: 0160-0753.

Behrendt, John Charles, and Wotorson, Cletus S. 1974. "Total-count gamma radiation map of the Gbanka Quadrangle, Liberia." U. S. Geological Survey, Reston, VA, United States. Miscellaneous Investigations Series Map. January 1, 1974. Report number: I-0776-C. Map Type: geophysical survey map. Annotation: Lat 6 degrees to 7 degrees, long 9 degrees to 10 degrees. Scale 1:250,000 (1 inch = about 4 miles). Sheet 26 by 29 inches. Descriptors: Africa; airborne; central; gamma ray; Gbanka Quadrangle; geophysical methods; geophysical surveys; Liberia; maps; radioactivity methods; surveys; USGS; West Africa. ISSN: 0160-0753.

Behrendt, John Charles and Wotorson, Cletus S. 1974. Total-count gamma radiation map of the Harper Quadrangle, Liberia. Miscellaneous Investigations Series Map. January 1, 1974. Report number: I-0780-C. U. S. Geological Survey, Reston, VA, United States. Map Type: geophysical survey map. Scale 1:250,000 (1 inch = about 4 miles). Sheet 29 by 38 inches. Descriptors: Africa; airborne; gamma ray; geophysical methods; geophysical surveys; Harper Quadrangle; Liberia; maps; radioactivity methods; southeast; surveys; USGS; West Africa. ISSN: 0160-0753.

Behrendt, John Charles and Wotorson, Cletus S. 1974. "Total-count gamma radiation map of the Juazohn Quadrangle, Liberia." U. S. Geological Survey, Reston, VA, United States. Miscellaneous Investigations Series Map. January 1, 1974. Report number: I-0779-C. Map Type: geophysical survey map. Scale 1:250,000 (1 inch = about 4 miles). Sheet 29 by 40 inches. Descriptors: Africa; airborne; gamma ray; geophysical methods; geophysical surveys; Juazohn Quadrangle; Liberia; maps; radioactivity methods; southeast; surveys; USGS; West Africa. ISSN: 0160-0753.

Behrendt, John Charles and Wotorson, Cletus S. 1974. "Total-count gamma radiation map of the Monrovia Quadrangle, Liberia." U. S. Geological Survey, Reston, VA, United States. Miscellaneous Investigations Series Map. January 1, 1974. Report number: I-0775-C. Map Type: geophysical survey map. Scale 1:250,000 (1 inch = about 4 miles). Sheet 29 by 36 inches. Descriptors: Africa; airborne; gamma ray; geophysical methods; geophysical surveys; Liberia; maps; Monrovia Quadrangle; radioactivity methods; surveys; USGS; West Africa. ISSN: 0160-0753.

Behrendt, John Charles and Wotorson, Cletus S. 1974. "Total-count gamma radiation map of the Sanokole Quadrangle, Liberia." U. S. Geological Survey, Reston, VA, United States. Miscellaneous Investigations Series Map. January 1, 1974. Report number: I-0774-C. Map Type: geophysical survey map. Scale 1:250,000 (1 inch = about 4 miles). Sheet 25 1/2 by 29 inches. Descriptors: Africa; airborne; gamma ray; geophysical methods; geophysical surveys; Liberia; maps; northeast; radioactivity methods; Sanokole Quadrangle; surveys; USGS; West Africa. ISSN: 0160-0753.

Behrendt, John Charles and Wotorson, Cletus S. 1974. "Total-count gamma radiation map of the Voinjama Quadrangle, Liberia." U. S. Geological Survey, Reston, VA, United States. Miscellaneous Investigations Series Map. January 1, 1974. Report number: I-0771-C. Map Type: geophysical survey map. Scale 1:250,000 (1 inch = about 4 miles). Sheet 29 by 32 inches. Descriptors: Africa; airborne; gamma ray; geophysical methods; geophysical surveys; Liberia; maps; north; radioactivity methods; surveys; USGS; Voinjama Quadrangle; West Africa. ISSN: 0160-0753.

Behrendt, John Charles and Wotorson, Cletus S. 1974. "Total-count gamma radiation map of the Zorzor Quadrangle, Liberia." U. S. Geological Survey, Reston, VA, United States. Miscellaneous Investigations Series Map. January 1, 1974. Report number: I-0773-C. Map Type: geophysical survey map. Annotation: Lat 7 degrees to 8 degrees, long 9 degrees to 10 degrees. Scale 1:250,000 (1 inch = about 4 miles). Sheet 26 by 29 inches. Descriptors: Africa; airborne; gamma ray; geophysical methods; geophysical surveys; Liberia; maps; north; radioactivity methods; surveys; USGS; West Africa; Zorzor Quadrangle. ISSN: 0160-0753.

Behrendt, John Charles and Wotorson, Cletus S. 1972. "Aeromagnetic and gravity investigations of the sedimentary basins on the continental shelf and coastal area of Liberia, West Africa." In: African Geology; Structural Geology. Pages 571-581. 1972. Univ. Ibadan, Dep. Geol., Ibadan, Nigeria. Descriptors: Africa; airborne; Atlantic Ocean; basins; coastal; continental shelf; geophysical methods; geophysical surveys; gravity methods; Liberia; magnetic methods; marine; sedimentary basins; surveys; tectonics; West Africa. Notes: With discussion. Illustrations, including sketch map.

Behrendt, John Charles and Wotorson, Cletus S. 1972. "Aeromagnetic and gravity investigations of the sedimentary basins on the continental shelf and coastal area of Liberia, west Africa." Special Papers - Liberia, Geological Survey. 2; 1972. Monrovia, Liberia. Pages: 13. Publication: Monrovia: Printed by the Printing Division of the Ministry of Information, Cultural Affairs, and Tourism. Descriptors: Africa; airborne; anomalies; coastal; continental shelf; geologic; geophysical methods; geophysical surveys; gravity methods; Liberia; magnetic methods; maps; marine geology; offshore; sedimentation; surveys; West; West Africa. Subjects: Geology- Liberia. Rocks, Sedimentary. Notes: Includes bibliographical references (p. 12-13). 13 pages: ill.; 28 cm. Map scale: 1:250,000. Geologic maps. Illustrations. ISSN: 0375-6831; OCLC: 2980872.

Behrendt, John Charles and Wotorson, Cletus S. 1972. "Tectonic map of Liberia based on geophysical and geological surveys." Open-File Report. Report number: OF 72-0032. Pages: 78. U. S. Geological Survey, Reston, VA, United States. January 1, 1972. Map Type: tectonic map. Illustrations. Notes: Cover title. "Project report, Liberian Investigations, IR-LI-60." Notes: Some of illustrative matter folded in pocket. Prepared under the auspices of the Agency for International Development, U.S. Dept. of State. Referred to in press release dated May 3, 1972. Series: IR-LI-60. Descriptors: Africa; explanatory text; geophysical surveys; Liberia; maps; structural geology; surveys; tectonic maps; USGS; West Africa. Contents: tectonic map of Liberia-interpretation from geophysical and geological data; total magnetic intensity; generalized residual total magnetic intensity; generalized total count Gamma radiation map; geologic contact inferred from Gamma radiation data; Bouguer gravity anomalies. Structural geology. ISSN: 0196-1497; OCLC: 25309005; USGS Library; Online: http://onlinepubs.er.usgs.gov/djvu/OFR/1972/ofr_72_32.djvu

Behrendt, John Charles and Wotorson, Cletus S. 1971. "An aeromagnetic and aeroradioactivity survey of Liberia, west Africa." Geophysics. 36; 3, Pages 590-604. 1971. Society of Exploration Geophysicists. Tulsa, OK, United States. Abstract: A 140,000 km aeromagnetic and total-count gamma radiometric survey was made over Liberia in 1967-68 along north-south lines spaced 0.8 km over land and 4 km over the continental shelf. The data approximately delineate the boundary between the Liberian (ca. 2700 m.y.) age province in the northwestern two-thirds of the country, and the Pan-African (ca. 550 m.y.) age province in the coastal area of the northwestern two-thirds of the country, as well as a boundary marking the northwest extent of the isoclinally folded paragneisses and migmatites deformed within the Eburnean (ca. 2000 m.y.) age province in the southeast one-third. A zone of diabase dikes about 90 km inland can be traced, parallel to the coast from Sierra Leone to Ivory Coast on the basis of the magnetic data. Another zone of diabase dikes about 185 m.y. old is located along the coastal area and beneath the continental shelf parallel to the coast northwest of Greenville. Intrusion of these dikes probably coincides with the separation of Africa from North and South America. The magnetic data suggest basins of sedimentary rocks possibly 5 km thick on the continental shelf. The map indicates high-amplitude magnetic anomalies greater than 600 gammas; some reach amplitudes as great as 18,000 gammas over iron formation and about 1800 gammas over mafic and ultramafic intrusive bodies. The radioactivity data have a background level less than 100 counts per second (cps) over mafic granulite-facies rocks and unmetamorphosed sedimentary rocks in the coastal area. Granitic rocks have the greatest variation. The central area of the country has the highest background radiation level with large areas above 250 cps; the level in the eastern one-third of the country is low. These data are proving quite useful in reconnaissance geologic mapping. All anomalies over 500 cps are shown; some reach amplitudes over 750 cps. Total-count radiation levels have a significant correlation with percent K (sub 2) O in bedrock analyses, but anomalous amounts of Th and U must be present to account for the highest amplitude anomalies. A few specific anomalies have been correlated with concentrations of monazite and zircon in bedrock as well as in beach deposits. Descriptors: Africa; airborne; geophysical methods; geophysical surveys; Liberia; magnetic; magnetic methods; radioactivity; radioactivity methods; regional; surveys; West Africa; Applied geophysics. Illustrations, including map. ISSN: 0016-8033.

Behrendt, John Charles and Wotorson, Cletus S. 1971. "Aeromagnetic and Gravity investigations of the sedimentary basins on the Continental Shelf and coastal area of Liberia, West Africa." In: Conference on African geology, 1st. Pages 4-5. 1971. Commonwealth Geological Liaison Office, London. Descriptors: Africa; coastal; continental shelf; geophysical methods; geophysical surveys; gravity methods; Liberia; magnetic methods; surveys; West Africa; Applied geophysics.

Behrendt, John Charles and Wotorson, Cletus S. 1971. "Aeromagnetic evidence that the Pan-African thermo tectonic event imprinted Liberian age metamorphic rocks in Liberia, West Africa." Abstracts with Programs - Geological Society of America. 3; 7, Pages 502. 1971. Geological Society of America (GSA). Boulder, CO, United States. Descriptors: Africa; geophysical methods; geophysical surveys; interpretation; Liberia; magnetic methods; regional; surveys; West Africa; Applied geophysics. ISSN: 0016-7592

Behrendt, John Charles and Wotorson, Cletus. 1971. "Aeromagnetic map of the Bopolu Quadrangle, Liberia." U. S. Geological Survey, Reston, VA, United States. Open-File Report. Report number: OF 71-0020. Pages: (2 sheets). January 1, 1971. Map Scale: 1:250,000. Map Type: magnetic survey map. Annotation: 2 sheets, scale 1:250,000 (1 inch = about 4 miles). Subjects: Africa; airborne methods; Bopolu Quadrangle; geophysical methods; geophysical surveys; Liberia; magnetic methods; maps; surveys; USGS; West Africa. ISSN: 0196-1497.

Behrendt, John Charles and Wotorson, Cletus S. 1971. "Aeromagnetic map of the Gbanka Quadrangle, Liberia." U. S. Geological Survey, Reston, VA, United States. Open-File Report. Report number: OF 71-0022. Pages: 5 (1 sheet). January 1, 1971. Map Scale: 1:250,000. Map Type: magnetic survey map. 5 [i.e. 8] leaves: color map; 27 cm. + map 44 x 45 cm. folded in envelope 30 x 24 cm. Annotation: 1 sheet, scale 1:250,000 (1 inch = about 4 miles). Descriptors: Africa; airborne methods; Gbanka Quadrangle; Geomagnetism; geophysical methods; geophysical surveys; Liberia; magnetic methods; maps; surveys; USGS; West Africa; maps. Notes: "Folio of the Gbanka quadrangle, Liberia." "Prepared under the joint sponsorship of the government of Liberia and the Agency for International Development, U.S. Department of State." Includes text with bibliography, index map, and 2 marginal maps, on reduced scale: Tectonic map, and Residual total magnetic intensity map. ISSN: 0196-1497; OCLC: 20872388.

Behrendt, John Charles and Wotorson, Cletus S. 1971. "Aeromagnetic map of the Harper Quadrangle, Liberia." U. S. Geological Survey, Reston, VA, United States. Open-File Report. Report number: OF 71-0023. Pages: 5 (3 sheets). January 1, 1971. Map Scale: 1:250,000. Map Type: magnetic survey map. Annotation: 3 sheets, scale 1:250,000 (1 inch = about 4 miles). Notes: Five folded maps in pocket. Referred to in press release dated August 2, 1971. Bibliography: leaf 5. Descriptors: Africa; airborne methods; geophysical methods; geophysical surveys; Harper Quadrangle; Liberia; magnetic methods; maps; surveys; USGS; West Africa. ISSN: 0196-1497.

Behrendt, John Charles and Wotorson, Cletus S. 1971. "Aeromagnetic map of the Monrovia Quadrangle, Liberia." U. S. Geological Survey, Reston, VA, United States. Open-File Report. Report number: OF 71-0024. Pages: 8 (2 sheets). January 1, 1971. Map Type: magnetic survey map. Annotation: 2 sheets, scale 1:250,000 (1 inch = about 4 miles). Notes: At head of Project report, Liberian investigations (IR) LI-66 C. Some maps folded in pocket. Referred to in press release dated August 2, 1971. Descriptors: Africa; airborne methods; geophysical methods; geophysical surveys; Liberia; magnetic methods; maps; Monrovia Quadrangle; surveys; USGS; West Africa. ISSN: 0196-1497.

Behrendt, John Charles and Wotorson, C. S. 1971. "Aeromagnetic map of the Sanokole Quadrangle, Liberia" Open-File Report. Report number: OF 71-0025. Pages: 6 (1 sheet). U. S. Geological Survey, Reston, VA, United States. January 1, 1971. Map Scale: 1:250,000. Map Type: magnetic survey map. Annotation: 1 sheet, scale 1:250,000 (1 inch = about 4 miles). Notes: At head of Project report, Liberian investigations (IR) LI-67 C. Some maps folded in pocket. Referred to in press release dated August 2, 1971. Bibliography: leaf 6. Descriptors: Africa; airborne methods; geophysical methods; geophysical surveys; Liberia; magnetic methods; maps; Sanokole Quadrangle; surveys; USGS; West Africa. ISSN: 0196-1497.

Behrendt, John Charles and Wotorson, C. S. 1971. "Aeromagnetic map of the Voinjama Quadrangle, Liberia." Open-File Report. Report number: OF 71-0026. Pages: 7 (2 sheets). U. S. Geological Survey, Reston, VA, United States. January 1, 1971. Map Scale: 1:250,000. Map Type: magnetic survey map. Annotation: 2 sheets, scale 1:250,000 (1 inch = about 4 miles). Notes: Project report, Liberian investigations (IR) LI-68 C. Referred to in press release dated August 2, 1971. Bibliography: leaves 6-7. Descriptors: Africa; airborne methods; geophysical methods; geophysical surveys; Liberia; magnetic methods; maps; surveys; USGS; Voinjama Quadrangle; West Africa. ISSN: 0196-1497.

Behrendt, John Charles and Wotorson, Cletus S. 1971. "Aeromagnetic map of the Zorzor Quadrangle, Liberia." U. S. Geological Survey, Reston, VA, United States. Open-File Report. Report number: OF 71-0027. Pages: 7 (1 sheet). January 1, 1971. Map Scale: 1:250,000. Map Type: magnetic survey map. Annotation: 1 sheet, scale 1:250,000 (1 inch = about 4 miles). Descriptors: Africa; airborne methods; geophysical methods; geophysical surveys; Liberia; magnetic methods; maps; surveys; USGS; West Africa; Zorzor Quadrangle. ISSN: 0196-1497.

Behrendt, John Charles and Wotorson, Cletus S. 1971. "Aeromagnetic survey of Liberia, West Africa, and its relation to the separation of Africa and South America." In: Ocean Floor Spreading, 4. International Union of Geodesy and Geophysics, 15th General Assembly, Moscow, 1971, Abstracts. Pages 29. 1971. Descriptors: Africa; airborne; anomalies; continental drift; geophysical methods; geophysical surveys; interpretation; Liberia; magnetic anomalies; magnetic methods; surveys; West Africa; Applied geophysics.

Behrendt, John Charles and Wotorson, Cletus S. 1971. "Bouguer anomaly map of the Monrovia Quadrangle, Liberia." Open-File Report. Report number: OF 71-0028. Pages: 5 (1 sheet). U. S. Geological Survey, Reston, VA, United States. January 1, 1971. Map Scale: 1:250,000. Map Type: gravity survey map. Annotation: 1 sheet, scale 1:250,000 (1 inch = about 4 miles). Descriptors: Africa; airborne methods; Bouguer anomalies; geophysical methods; geophysical surveys; gravity anomalies; gravity methods; Liberia; maps; Monrovia Quadrangle; surveys; USGS; West Africa. Referred to in press release dated Oct. 12, 1971. ISSN: 0196-1497.

Behrendt, John Charles and Wotorson, Cletus S. 1971. "Results of regional gravity survey of Liberia, west Africa." Eos, Transactions, American Geophysical Union. 52; 4, Pages 182. 1971. American Geophysical Union. Washington, DC, United States. Descriptors: Africa; geophysical methods; geophysical surveys; gravity methods; Liberia; regional; surveys; West Africa; applied geophysics. ISSN: 0096-3941.

Behrendt, John Charles and Wotorson, Cletus S. 1971. "Total-count gamma radiation map of the Bopolu Quadrangle, Liberia." Open-File Report. Report number: OF 71-0029. Pages: 5 (2 sheets). U. S. Geological Survey, Reston, VA, United States. January 1, 1971. Map Scale: 1:250,000. Map Type: geophysical survey map. Annotation: 2 sheets, scale 1:250,000 (1 inch = about 4 miles). Notes: Referred to in press release dated August 16, 1971. Four maps folded in pocket. Bibliography: leaf 5. Descriptors: Africa; airborne methods; Bopolu Quadrangle; geophysical methods; geophysical surveys; Liberia; maps; radioactivity methods; surveys; USGS; West Africa. ISSN: 0196-1497.

Behrendt, John Charles and Wotorson, Cletus S. 1971. "Total-count gamma radiation map of the Buchanan Quadrangle, Liberia." Open-File Report. Report number: OF 71-0030. Pages: 5 (2 sheets). U. S. Geological Survey, Reston, VA, United States. January 1, 1971. Map Scale: 1:250,000. Map Type: geophysical survey map. Annotation: 2 sheets, scale 1:250,000 (1 inch = about 4 miles). Notes: Four folded maps in pocket. Referred to in press release dated Oct. 12, 1971. Bibliography: leaf 5. Descriptors: Africa; airborne methods; Buchanan Quadrangle; geophysical methods; geophysical surveys; Liberia; maps; radioactivity methods; surveys; USGS; West Africa. ISSN: 0196-1497.

Behrendt, John Charles and Wotorson, Cletus S. 1971. "Total-count gamma radiation map of the Gbanka Quadrangle, Liberia." Open-File Report. Report number: OF 71-0031. Pages: 5 (1 sheet). U. S. Geological Survey, Reston, VA, United States. January 1, 1971. Map Scale: 1:250,000. Map Type: geophysical survey map. Annotation: 1 sheet, scale 1:250,000 (1 inch = about 4 miles). Notes: Two folded maps in pockets. Referred to in press release dated Oct. 12, 1971. Bibliography: leaf 5. Descriptors: Africa; airborne methods; Gbanka Quadrangle; geophysical methods; geophysical surveys; Liberia; maps; radioactivity methods; surveys; USGS; West Africa. ISSN: 0196-1497.

Behrendt, John Charles and Wotorson, Cletus S. 1971. "Total-count gamma radiation map of the Harper Quadrangle, Liberia." Open-File Report. Report number: OF 71-0032. Pages: 5 (3 sheets). U. S. Geological Survey, Reston, VA, United States. January 1, 1971. Map Scale: 1:250,000. Map Type: geophysical survey map. Annotation: 3 sheets, scale 1:250,000 (1 inch = about 4 miles). Notes: Six folded maps in pocket. Referred to in press release dated Oct. 12, 1971. Bibliography: leaf 5. Descriptors: Africa; airborne methods; geophysical methods; geophysical surveys; Harper Quadrangle; Liberia; maps; radioactivity methods; surveys; USGS; West Africa. ISSN: 0196-1497.

Behrendt, John Charles and Wotorson, Cletus S. 1971. "Total-count gamma radiation map of the Juazohn Quadrangle, Liberia." Open-File Report. Report number: OF 71-0033. Pages: 6 (1 sheet). U. S. Geological Survey, Reston, VA, United States. January 1, 1971. Map Scale: 1:250,000. Map Type: geophysical survey map. Annotation: 2 sheets, scale 1:250,000 (1 inch = about 4 miles). Notes: Four folded maps in pocket. Referred to in press release dated Oct. 12, 1971. Bibliography: leaf 6. Descriptors: Africa; airborne methods; geophysical methods; geophysical surveys; Juazohn; Liberia; maps; radioactivity methods; surveys; USGS; West Africa. ISSN: 0196-1497.

Behrendt, John Charles and Wotorson, Cletus S. 1971. "Total-count gamma radiation map of the Monrovia quadrangle, Liberia." Washington: U.S. Geological Survey, 1971. 6 leaves: maps; 27 cm. Open-file report (Geological Survey (U.S.)); 1608. Notes: Four maps folded in pocket. Referred to in press release dated August 16, 1971. Bibliography: leaf 6. Descriptors: Gamma rays; Measurement. USGS Library.

Behrendt, John Charles and Wotorson, Cletus S. 1971. "Total-count gamma radiation map of the Monrovia Quadrangle, Liberia." Open-File Report. Report number: OF 71-0034. Pages: 6 (2 sheets). U. S. Geological Survey, Reston, VA, United States. January 1, 1971. Map Scale: 1:250,000. Map Type: geophysical survey map. Annotation: 2 sheets, scale 1:250,000 (1 inch = about 4 miles). Descriptors: Africa; airborne methods; geophysical methods; geophysical surveys; Liberia; maps; Monrovia Quadrangle; radioactivity methods; surveys; USGS; West Africa. ISSN: 0196-1497.

Behrendt, John Charles and Wotorson, Cletus S. 1971. Total-count gamma radiation map of the Sanokole Quadrangle, Liberia. Open-File Report. Report number: OF 71-0035. Pages: 5 (1 sheet). U. S. Geological Survey, Reston, VA, United States. January 1, 1971. Map Type: geophysical survey map. Annotation: 2 sheets, scale 1:250,000 (1 inch = about 4 miles). Notes: Two maps folded in pocket. Referred to in press release dated August 16, 1971. Bibliography: leaf 5. Descriptors: Africa; airborne methods; geophysical methods; geophysical surveys; Liberia; maps; radioactivity methods; Sanokole Quadrangle; surveys; USGS; West Africa. ISSN: 0196-1497.

Behrendt, John Charles and Wotorson, Cletus S. 1971. "Total-count gamma radiation map of the Voinjama Quadrangle, Liberia." Open-File Report. Report number: OF 71-0036. Pages: 5 (2 sheets). U. S. Geological Survey, Reston, VA, United States. January 1, 1971. Map Scale: 1:250,000. Map Type: geophysical survey map. Annotation: 2 sheets, scale 1:250,000 (1 inch = about 4 miles). Notes: Four maps folded in pocket. Referred to in press release dated August 16,

1971. Bibliography: leaf 5. Descriptors: Africa; airborne methods; geophysical methods; geophysical surveys; Liberia; maps; radioactivity methods; surveys; USGS; Voinjama Quadrangle; West Africa. ISSN: 0196-1497.

Behrendt, John Charles and Wotorson, Cletus S. 1971. Total-count gamma radiation map of the Zorzor Quadrangle, Liberia. Open-File Report. Report number: OF 71-0037. Pages: 6 (1 sheet). U. S. Geological Survey, Reston, VA, United States. January 1, 1971. Map Type: geophysical survey map. Annotation: 1 sheet, scale 1:250,000 (1 inch = about 4 miles). Notes: Two folded maps in pocket. Referred to in press release dated August 16, 1971. Bibliography: leaf 6. Descriptors: Africa; airborne methods; geophysical methods; geophysical surveys; Liberia; maps; radioactivity methods; surveys; USGS; West Africa; Zorzor Quadrangle. ISSN: 0196-1497.

Behrendt, John Charles and Wotorson, Cletus S. 1971. "The use of aeromagnetic and aeroradioactivity surveys for geologic mapping in Liberia, West Africa." In: International Symposium on Remote Sensing of Environment, 7th, Proc., Vol. 3. Proceedings of the International Symposium on Remote Sensing of Environment. 7, Vol. 3; Pages 2133-2153. 1971. Environmental Research Institute of Michigan. Ann Arbor, MI, United States. Descriptors: Africa; areal geology; cartography; geologic; geophysical methods; geophysical surveys; interpretation; Liberia; magnetic methods; radioactivity methods; regional; remote sensing; surveys; West Africa. Illustrations including sketch maps. ISSN: 0275-5505.

Behrendt, John Charles and Wotorson, Cletus S. 1970. "Aeromagnetic and gravity investigations of the coastal area and continental shelf of Liberia, West Africa, and their relation to continental drift." Geological Society of America Bulletin. 81; 12, Pages 3563-3573. 1970. Abstract: Sequence of tectonic events suggested. Descriptors: Africa; Atlantic Ocean; coast; continental shelf; crust; crustal structure; geophysical methods; geophysical surveys; gravity; gravity methods; Liberia; magnetic; magnetic methods; structure; surveys; tectonics; West Africa; applied geophysics. Illustrations, including map. ISSN: 0016-7606

Behrendt, John Charles and Wotorson, Cletus S. 1970. "Aeromagnetic map of the Buchanan Quadrangle, Liberia." Open-File Report. Report number: OF 71-0021. Pages: (2 sheets). U. S. Geological Survey, Reston, VA, United States. January 1, 1971. Scale: 1:250,000. Map Type: magnetic survey map. 5 [i.e. 8] leaves: maps; 27 cm. Annotation: 2 sheets, scale 1:250,000 (1 inch = about 4 miles). Descriptors: Africa; airborne methods; Buchanan Quadrangle; geophysical methods; geophysical surveys; Liberia; magnetic methods; maps; surveys; USGS; West Africa. Notes: Four folded maps in pocket. Referred to in press release dated August 2, 1971. Bibliography: leaf 5. ISSN: 0196-1497

Behrendt, John Charles and Wotorson, Cletus S. 1970. "High-amplitude radioactivity anomalies in Liberia." Reston, VA, United States: U.S. Geological Survey. 1 map. Scale 1: 1,000,000. Open-file report (Geological Survey (U.S.)); 70-23. Descriptors: Radioactive substances; Liberia. USGS Library.

Behrendt, John Charles and Wotorson, Cletus S. 1970. "Total intensity aeromagnetic map of Liberia, including the continental shelf." Reston, VA, United States: U.S. Geological Survey. 2

maps. Scale 1: 1,000,000. Open-file report (Geological Survey (U.S.)); 70-24. Descriptors: Continental shelf; Liberia. USGS Library.

Behrendt, John Charles and Wotorson, Cletus. 1969. "Aeromagnetic, aeroradioactivity and gravity surveys in Liberia, west Africa." Eos, Transactions, American Geophysical Union. 50; 4, Pages 320. 1969. American Geophysical Union. Washington, DC, United States. Descriptors: aeromagnetic and aeroradioactivity gravity; Africa; airborne; geophysical methods; geophysical surveys; gravity methods; Liberia; magnetic methods; radioactivity methods; regional; surveys; West Africa; applied geophysics. ISSN: 0096-3941

Behrendt, John Charles and Wotorson, Cletus S. 1969. "Aeromagnetic and gravity investigations of the sedimentary basins on the continental shelf and coastal area of Liberia, West Africa." Washington: U.S. Geological Survey; Liberian Geological Survey; United States. Agency for International Development. 11 [i.e. 17] leaves: ill., maps; 27 cm. Series: Open-file report (Geological Survey (U.S.)); 1327. Notes: Prepared as part of the joint Geological Exploration and Resources Appraisal project of the Liberian Geological Survey and the U.S. Geological Survey sponsored by the Government of Liberia and United States Agency for International Development. Four maps folded in pocket. Referred to in press release dated November 10, 1969. Bibliography: leaf 11. Descriptors: Gravity; Liberia. USGS Library.

Behrendt, John Charles and Wotorson, Cletus S. 1969. "Preliminary Bouguer anomaly map of the Monrovia region of Liberia." Reston, VA, United States: U.S. Geological Survey. 1 map. Scale 1: 500,000. Open-file report (Geological Survey (U.S.)); 69-17. USGS Library.

Behrendt, John Charles and Wotorson, Cletus S. 1969. "Preliminary magnetic basement elevation of Liberian Continental Shelf." Reston, VA, United States: U.S. Geological Survey. 1 map. Scale 1: 1,000,000. Open-file report (Geological Survey (U.S.)); 69-18. Descriptors: Continental shelf; Liberia; Maps. USGS Library.

Behrendt, John Charles ; Schlee, John and Robb, James M. 1973. "Magnetic and Gravity Investigations across the Edge of the African Continental Crust on the Liberian Continental Margin." Eos, Transactions, American Geophysical Union. 54; 4, Page 332. 1973. American Geophysical Union. Washington, DC. Descriptors: Africa; Atlantic Ocean; continental margin; continental slope; crust; distribution; fractures; geophysical methods; geophysical surveys; gravity methods; Liberia; magnetic methods; marine; structure; surveys; systems; West Africa; Solid earth geophysics. ISSN: 0096-3941.

Behrendt, John Charles ; Schlee, John S. and Robb, James M. 1972. "Magnetic anomalies on the continental margin of Africa off Liberia observed on USGS-IDOE Cruise Leg 5." Eos, Transactions, American Geophysical Union. 53; 4, Pages 408. 1972. American Geophysical Union. Washington, DC, United States. Descriptors: Africa; continental shelf; continental slope; geophysical methods; geophysical surveys; Liberia; magnetic methods; surveys; West Africa; Applied geophysics. ISSN: 0096-3941.

Behrendt, John C; Schlee, John; Robb, James M. and Silverstein, M. Katherine. 1974. "Structure of the Continental Margin of Liberia, West Africa." Geological Society of America Bulletin. 85; 7,

Pages 1143-1158. 1974. Descriptors: acoustical methods; Africa; bathymetry; coastal; continental margin; continental shelf; continental slope; crust; distribution; extent; faults; fracture zones; geophysical methods; geophysical surveys; gravity methods; history; interpretation; Liberia; magnetic methods; maps; marine geology; sea floor spreading; structure; surveys; tectonics; tectonophysics; West Africa; solid earth geophysics. Illustrations, including sketch maps. ISSN: 0016-7606.

Bennett, Kara C. and Rusk, Don. 2002. Regional 2D seismic interpretation and exploration potential of offshore deepwater Sierra Leone and Liberia, West Africa. In: Sternbach, Linda R. (editor); Tari, Gabor (editor). West Africa; a special issue on an emerging world petroleum province. The Leading Edge (Tulsa, OK). 21; 11, Pages 1118-1124. 2002. Society of Exploration Geophysicists. Tulsa, OK, United States. 2002. Descriptors: Africa; Atlantic Ocean; bathymetry; block structures; Buchanan fault zone; continental shelf; correlation; deep water environment; East Atlantic; fault zones; faults; genesis; geophysical methods; geophysical profiles; geophysical surveys; Greenville fault zone; Harper Basin; Liberia; Liberia Basin; mapping; natural gas; offshore; petroleum; petroleum exploration; reconstruction; reservoir-rocks; sea floor spreading; seismic methods; seismic profiles; Sierra Leone; Sierra Leone Basin; source rocks; structural traps; surveys; systems; tectonics; traps; two dimensional models; unconformities; West Africa. Illustrations: sections, sketch maps. ISSN: 1070-485X.

Bennett, K. C; Rusk, D. C. and Mohn, K. W. 2002. "Structural framework and potential hydrocarbon plays in offshore Sierra Leone and Liberia." Conference Sponsor: Enterprise Oil, AGIP: Exploration and Production Division. Conference: 64th Annual Meeting of the European Association of Geoscientists and Engineers, Florence (Italy), 27-30 May 2002. (World Meeting Number 0005907). Notes: European Association of Geoscientists and Engineers, P.O. Box 593990 DB Houten, The Netherlands; URL: www.eage.nl. Paper No. G008.

Berge, John W. 1967. "Contributions to the petrology of the Goes Range area, Grand Bassa County, Liberia." Bulletin of the Geological Institutions of the University of Uppsala, New Series. 43; 4-5, 1967. University of Uppsala, Institute of Geology. Uppsala, Sweden. Pages: 24. Abstract: Two mutually conformable groups of Archaean rocks underlying the Goe Range area are described. The first, consisting mainly of felsic syenite gneisses, with minor bands of amphibolites is termed Archaean undefined. The second, called the Goe Range series, consists of metasedimentary rocks, in which argillaceous schists, quartzites, and iron formation are predominant. Petrographic and chemical evidence indicate that chemical precipitation and deposition of finely clastic rocks alternated during the period of sedimentation. Descriptors: Africa; Goe range area; Liberia; metamorphic rocks; petrology; Precambrian; West Africa; Igneous and metamorphic petrology. Map Scale: 1:40,000. Illustrations, including geol. Map. ISSN: 0302-2749.

Berge, John W. 1966. "Genetical aspects of the Nimba iron ores." Bulletin of the Geological, Mining and Metallurgical Society of Liberia. 1; 1, Pages 36-43. 1966. Geological, Mining and Metallurgical Society. Monrovia, Liberia. Abstract: Precambrian itabirite iron formation, hematite and goethite-hematite ore features, various theories presented, Liberia. Descriptors: Africa; economic geology; genesis; iron; Liberia; metals; mineral deposits, genesis; Nimba Gbahm ridges; properties; reserves; structure; West Africa; Economic geology of ore deposits. ISSN: 0367-4819.

Berge, John W. 1974. "Geology, Geochemistry, and Origin of the Nimba Itabirite and Associated Rocks, Nimba County, Liberia." Economic Geology and the Bulletin of the Society of Economic Geologists. 69; 1, Pages 80-92. 1974. Economic Geology Publishing Company. Lancaster, PA, United States. Descriptors: Africa; chemically precipitated rocks; clay minerals; economic geology; epidote amphibolite facies; facies; foliation; Gbahm Ridge Formation; geochemistry; hematite; iron formations; iron ores; iron rich rocks; itabirite; kaolinite; Liberia; magnetite; metal ores; metamorphic rocks; metamorphism; mineral deposits, genesis; mineralogy; Mount Alpha Formation; Mount Gbahm; Nimba County; Nimba Mountains; Nimba Supergroup; ore deposits; oxides; phyllites; Precambrian; processes; sedimentary rocks; sheet-silicates; silicates; weathering; West Africa; Economic geology of ore deposits; geological sketch maps. ISSN: 0361-0128.

Berge, John W. 1973. "The geology and origin of the Precambrian Goe Range iron formation and associated metasediments." Geologiska Foereningen i Stockholm Foerhandlingar. 95, Part 4; 555, Pages 363-373. 1973. Geological Society of Sweden. Stockholm, Sweden. Abstract: Geochemistry, genesis, Liberia. Descriptors: Africa; aluminum ores; chemically precipitated rocks; chert; complexes; genesis; geochemistry; gneisses; Goe Range; Goe Range Series; igneous rocks; iron formations; iron ores; Liberia; metachert; metal ores; metamorphic rocks; metasedimentary rocks; ore deposits; orogeny; petrology; phosphorus; plutonic rocks; Precambrian; quartzites; schists; sedimentary rocks; sedimentation; syenites; weathering; West Africa; Igneous and metamorphic petrology. Illustrations, including sketch maps. ISSN: 0016-786X.

Berge, John W. 1971. "Iron formation and supergene iron ores of the Goe range area, Liberia." Economic Geology and the Bulletin of the Society of Economic Geologists. 66; 6, Pages 947-960. 1971. Economic Geology Publishing Company. Lancaster, PA, United States. Descriptors: Africa; economic geology; genesis; geochemistry; Goe-range; iron; Liberia; metals; mineral-deposits,-genesis; mineralogy; West Africa; Economic geology of ore deposits. Illustrations, including sketch maps. ISSN: 0361-0128.

Berge, John W. 1970. "Implications of phosphorous fixation during chemical weathering in the genesis of super-gene iron ores." Bulletin - Geological, Mining and Metallurgical Society of Liberia. 4; Pages 33-43. 1970. Notes: Vol. 4. Geological, Mining and Metallurgical Society. Monrovia, Liberia. Descriptors: Africa; Bambuta; Bomi Hills; economic geology; genesis; iron ores; Liberia; metal ores; mineral deposits; genesis; ore deposits; processes; supergene-processes; weathering; West Africa. ISSN: 0367-4819.

Berge, John W. 1954. "General Geology of the Nimba Iron Ore Mine." The Liberian Naturalist. Volume 2, no. 3, September 1954, pages 16-19.

Berge, J. W.; Johansson, K. and Jack, J. 1977. "Geology and origin of the hematite ores of the Nimba Range, Liberia." Economic Geology and the Bulletin of the Society of Economic Geologists. 72; 4, Pages 582-607. 1977. Economic Geology Publishing Company. Lancaster, PA. Descriptors: Africa; areal geology; classification; diagenesis; economic geology; genesis; grade; hematite; hydrothermal alteration; hydrothermal processes; iron ores; leaching; Liberia; metal ores; metamorphism; metasomatism; mineral deposits, genesis; Nimba Mountains; ore bodies; ore deposits; oxides; petrography; processes; structural controls; structure; supergene processes;

syngenesis; West Africa; economic geology of ore deposits. References: 25; illustrations, including tables, geol. sketch map. ISSN: 0361-0128.

Bernasconi, A. 1987. "The major Precambrian terranes of eastern South America; a study of their regional and chronological evolution." Precambrian Research. 37; 2, Pages 107-124. 1987. Elsevier. Amsterdam, International. Descriptors: Africa; Amazonian Province; eastern South America; Liberia; Limpopo; Limpopo Belt; Paramirim Province; Precambrian; Ribeira Belt; Rio de la Plata Province; Sao Luis Province; South America; stratigraphy; Uruacuano Belt; West Africa. References: 76; illustrations, including 1 table, geol. sketch map. ISSN: 0301-9268.

Bessoles, B. 1977. "Geologie de l'Afrique; le craton ouest Africain." Translated title: "Geology of Africa; the West African Craton." Memoires du B.R.G.M.. 88, 1977. Bureau de Recherches Geologiques et Minières, (BRGM). Paris, France. Pages: 402. Language: French; Summary Language: English. Descriptors: absolute age; Africa; Algeria; areal geology; Birrimian; Burkina Faso; correlation; cratons; extent; geochronology; geologic maps; Guinea; igneous rocks; Ivory Coast; Kayes window; Kenieba window; Liberia; Liberian; Mali; Man Ridge; maps; Mauritania; metamorphic-rocks; metamorphism; Niger; North Africa; orogeny; Paleoproterozoic; Pan African Orogeny; petrology; Precambrian; Proterozoic; Reguibat Ridge; Sahara; Sierra Leone; stratigraphy; structural-geology; tectonics; upper Precambrian; West Africa; West African Shield. References: 392; illustrations, including tables, chart, strat. color, sects. ISSN: 0071-8246.

Beuttikofer, Johann. 1890. Reisebilder aus Liberia: Resultate Geographischer, Naturwissenschaftlicher und Ethnographischer. Untersuchungen weahrend der Jahre 1879-1882 und 1886-1887 von J. Beuttikofer. Translated "Travel pictures from Liberia: Results geographic, scientific and ethnographic. Investigations for the years 1879-1882 und 1886-1887." Leiden: E. J. Brill, 1890. Description: Book; Map. 2 v.: ill., maps (1 folded in pocket), portraits, plates; 24 cm. Contents Notes: V. 1. Reise- und Charakterbilder. -- v. 2. Die Bewohner Liberia's. - Thierwelt. Notes: Some of the colored plates accompanied by guard sheets with descriptive letterpress. Includes bibliographical references and index. Descriptors: Natural history; Liberia; Liberia travel. University of Wisconsin, American Geographical Society Library – Rare Books Collection. Call Number: DT625 .B94 1890. Notes: Map missing.

Beyer, M. 1960. "Geolog pa Nimba." Translated title: "Geology of Nimba." Stockholm: Malm. Volume 9, page 10.

Billa, M.; Feybesse, J. L.; Bronner, G. and Milesi, J. P. 2000. "BIF of the Nimba and Simandou Ranges; units tectonically stacked onto an Archean basement during the Eburnian Orogeny." In: Brazil 2000; 31st international geological congress; abstracts volume. International Geological Congress, Abstracts = Congres Geologique International, Resumes. 31; Pages; unpaginated. 2000. Conference: Brazil 2000; 31st international geological congress. Rio de Janeiro, Brazil. August 6-17, 2000. Language: English. Descriptors: Africa; Archean; banded-iron formations; basement; Birimian; chemically precipitated rocks; chronology; Eburnean Orogeny; Guinea; igneous rocks; iron formations; iron ores; Ivory Coast; Liberia; Liberian Shield; lithostratigraphy; Man Shield; metal ores; metamorphic rocks; mineral exploration; Nimba Mountains; Paleoproterozoic; Precambrian; Proterozoic; sedimentary rocks; Simandou Mountains; stratigraphic units; tectonics; upper Precambrian; volcanic rocks; West Africa. Media: compact disc. CODEN: IGABBY.

Billa, M; Feybesse, JL; Bronner, G; Lerouge, C; Milesi, JP; Traore, S; Diaby, S. 1999. "Les Formations a quartzites rubanes ferrugineux des monts Nimba et du Simandou; des unites empilees tectoniquement, sur un "soubassement" plutonique Archeen (craton de Kenema-Man), lors de l'orogene Eburneen. » Translated "Banded ferruginous quartzite formations of the Nimba and Simandou ranges; tectonically stacked units on an Archean plutonic sub-basement (Kenema-Man Craton), during the Eburnean Orogeny." Comptes Rendus de l'Academie des Sciences, Serie II. Sciences de la Terre et des Planetes. 329; 4, Pages 287-294. 1999. Gauthier-Villars. Montrouge, France. 1999. Language: French. Summary Language: English. Abstract: The volcano-sediments of the Nimba and Simandou Ranges lie in tectonic contact on an Archean plutonic substratum. This contact is associated with tangential tectonics assigned to the Paleoproterozoic, which has caused a thickening of the upper crust by tectonic redoublement of the volcano-sedimentary formation. Further, the volcano-sediments were deposited between 2.615 and 2.25 Ga, on a previously metamorphosed substratum (around 2.8-2.72 Ga). This leads us 1) to consider that the tectonic contact corresponds pro parte to a stratigraphic discordance caught up in the tangential tectonism and 2) to question to which cycle (Archean or Paleoproterozoic?) the volcano-sediments belong. Descriptors: Africa; Archean; basement; compression-tectonics; cross sections; crustal thickening; ferruginous quartzite; Guinea; Ivory Coast; Kenema-Man-Craton; Liberia; lower Proterozoic; metamorphic rocks; metamorphism; Nimba Mountains; Pan-African Orogeny; Precambrian; Proterozoic; quartzites; Simandou Mountains; structural analysis; tectonics; unconformities; upper Precambrian; West Africa; West African Shield; Igneous and metamorphic petrology; structural geology. References: 17. References include data from PASCAL, Institute de l'Information Scientifique et Technique, Vandoeuvre les Nancy, France. ISSN: 1251-8050.

Black, Russell. 2003. "Ethical codes in humanitarian emergencies: From practice to research?" Disasters 27, no. 2 (2003) p. 95-108. Abstract: Notable strides have been made in recent years to develop codes of conduct for humanitarian intervention in conflicts on the part of international NGOs and UN organisations. Yet engagement by the academic and broader research communities with humanitarian crises and ongoing complex political emergencies remains relatively ad hoc and unregulated beyond the basic ethical guidelines and norms developed within universities for research in general, and within the governing and representative bodies of particular academic disciplines. This paper draws on a case study of research on humanitarian assistance to Liberia during that country's civil war from 1989 to 1996. The difficulties faced by humanitarian agencies in Liberia led to the development of two key sets of ethical guidelines for humanitarian intervention: the Joint Policy of Operations (JPO) and Principles and Policies of Humanitarian Operations (PPHO). This paper seeks to address what lessons, if any, these ethical guidelines, together with different experiences of conducting research in war-torn Liberia, can provide in terms of the role of academic researchers - and research itself - in humanitarian crises. References: 60 Descriptors: Hazards and Disaster Planning; International Aid and Investment; disaster relief. ISSN: 0361-3666.

Black, Russell. 1980. "Precambrian of West Africa." Episodes. 1980; 4, Pages 3-8. 1980. International Union of Geological Sciences (IUGS). Ottawa, ON, Canada. IGCP (International Geological Correlation Programme). Descriptors: Africa; Archean; displacements; economic geology; faults; Guinea; Ivory Coast; Liberia; mineral resources; Nigeria; Pan African Orogeny; Precambrian; Proterozoic; Sierra Leone; stratigraphy; thrust faults; upper Precambrian; West

Africa; West African-Shield. Notes: IGCP Project No. 164. References: 11; illustrations, including geol. sketch map. ISSN: 0705-3797.

Blade, Lawrence Vernon. 1970. "Geology of the Bushrod Island-New Georgia clay deposit near Monrovia, Liberia." Liberian Geological Survey; United States. Agency for International Development. Other Titles: Bushrod Island-New Georgia clay deposit near Monrovia, Liberia. Washington: U.S. Geological Survey. 35 leaves: ill., maps; 27 cm. Open-file report (Geological Survey (U.S.)); 1346. Notes: Part of the Geological Exploration and Resources Appraisal Project, a cooperative effort of the Government of Liberia and the U.S. Agency for International Development. Referred to in press release dated January 22, 1970. Bibliography: leaf 25. Descriptors: Clay, Liberia- Monrovia region; Borings, Liberia- Monrovia region. USGS LIBRARY.

Blay, T. O. 1980. "Report on Investigations of the Occurrence of Gold around Tutubli, Grand Bassa, County, Mosaic Blocks K-23, 24." Liberian Geological Survey unpublished un-numbered report, 26 pages. AMRS, Inc. See: http://www.africaminerals.com/

Boadi, Isaac Opoku. 1991. Origin of mega-gold placer deposits in the light of data on the Bukon Jedeh deposit, Liberia, and on the Tarkwa deposit, Ghana. Description: xiii, 220 leaves: color ill., maps (some color); 28 cm. Subjects: Gold ores- Geology- Liberia. Gold ores- Geology- Ghana. Placer deposits- Liberia. Placer deposits- Ghana. Descriptors: absolute-age; Africa; bedrock; Bukon Jedeh Deposit; chemical weathering; dates; disseminated deposits; economic geology; Ghana; gold ores; host-rocks; isochrons; Liberia; metal ores; ore grade; petrography; placers; Rb-Sr; Tarkwa-Deposit; weathering; West Africa. Notes: Includes bibliographical references (leaves 94-99, 208-213). Doctoral Thesis, New Mexico Institute of Mining and Technology, 1991. Abstract (Document Summary): The origin of mega-gold placer deposits is addressed. Field, petrographic and geochemical data are presented on a Cenozoic placer deposit in Liberia, and on a Proterozoic paleoplacer deposit in Ghana. The Bukon Jedeh deposit in Liberia is a major placer deposit with dimensions exceeding 50 sq. km. The deposit formed in situ by deep chemical weathering of underlying graphite-, pyroxene- and garnet bearing gneisses and gold occurs as nuggets that decrease in size downward. Gold grades range from 0.05 to 6 ppm. Trace element data on gold nuggets indicate they were formed by precipitation from ground waters. Assays of bedrock samples from the area reveal that the protore of this deposit is a low grade, broadly disseminated mineralization with grade ranging from 0.04 to 8 ppm. Petrographic and geochemical data on bedrock samples indicate their protoliths were part of a greenstone succession emplaced in a volcanic arc setting. The Tarkwa deposit in Ghana is hosted by immature and mature fluvial sedimentary units within the Tarkwa basin. Gold is dispersed through the immature sediments at grades of 0.1 to 2 ppm. In the mature sediments, gold is concentrated in well defined conglomerate bands at grades of 9 ppm. Volcanic fragments from the immature sediments have elevated concentrations of gold ranging from 0.05 to 5.7 ppm. The fragments are aluminous and have trace element chemistry that indicates dacite to rhyodacite protoliths in contrast to the Birimian metavolcanic and volcaniclastic rocks that have basaltic and andesitic protoliths. A Rb-Sr isochron age of 2.15 pm 0.11 Ga and an initial $sp\{87\}Sr$ $sp\{86\}Sr$ ratio of 0.70057 pm 0.43 were obtained for samples of Birimian metavolcanic rocks. Isotopic data on the volcanic fragments are similar to that on the Birimian metavolcanic rocks, and support derivation of sedimentary units in the Tarkwa basin and their gold content from a more felsic component of the Birimian greenstone succession with broadly disseminated primary mineralization. Results of this study show that favorable conditions for the formation of mega- gold placers are basins adjacent to terranes with regional scale disseminated mineralization, and a paleoclimatic condition, at time of sedimentation, favoring deep chemical weathering. Enriched soils and regolith mantle over such terranes erode into adjacent basins during uplift. The high levels of gold in mega-placers result from extensive reworking of enormous quantities of the low-grade auriferous sediments. OCLC: 26707247.

Boadi, Issac Opoku and Norman, David I. 1991. Formation of gold nuggets in laterite soils, Bukon Jedeh, Liberia, West Africa. In: Geological Society of America, 1990 annual meeting. Abstracts with Programs - Geological Society of America. 22; 7, Pages 43. 1990. Geological Society of America (GSA). Boulder, CO, United States. 1990. Conference: Geological Society of America, 1990 annual meeting. Dallas, TX, United States. Oct. 29-Nov. 1, 1990. Descriptors: Africa; Bukon-Jedeh; economic geology; gold ores; laterites; Liberia; metal ores; mineral-deposits,-genesis; mineralization; nuggets; placers; soils; West Africa; economic geology; geology-of-energy-sources. ISSN: 0016-7592.

Boadi, Issac Opoku and Norman, David I. 1990. "Gold mineralization in the Bukon Jedeh district of Liberia." In: 15th colloquium of African geology--15 (super e) colloque de geologie africaine. Publication Occasionnelle - Centre International Pour la Formation et les Echanges Geologiques = Occasional Publication - International Center for Training and Exchanges in the Geosciences. 20; Pages 338. 1990. Centre International pour la Formation et les Echanges Geologiques (CIFEG). Paris, France. 1990. Conference: 15th colloquium of African geology--15 (super e) colloque de geologie africaine. Nancy, France. Sept. 10-13, 1990. Descriptors: Africa; Bukon Jedeh Deposit; Bukon Jedeh mining district; gold ores; Liberia; metal ores; West Africa; Economic geology; geology of ore deposits. ISSN: 0769-0541.

Böckh, E. 1964. "Itabiritische Eisenerze in Liberia." Schr. Gesell. Deutscher Metallhütten und Bergleute. Number 14, 1964, pages 23-28.

Bong Mining Company. 1984. "Bong Mining Company Technical Information." Bong Mining Company, Monrovia, Liberia, 27 pages. AMRS, Inc. See: http://www.africaminerals.com/

Bonvalot, S.; Villeneuve, M.; Legeley, A. and Albouy, Y. 1988. "Leve gravimetrique du Sud-Ouest du craton Ouest-Africain. Translated "Gravimetric survey of the southwestern part of the West African Craton." Comptes Rendus de l'Academie des Sciences, Serie 2, Mecanique, Physique, Chimie, Sciences de l'Univers, Sciences de la Terre. 307; 18, Pages 1863-1868. Elsevier. Paris, France. Language: French; Summary Language: English. Descriptors: Africa; Bouguer anomalies; Gambia; geophysical methods; geophysical surveys; gravity anomalies; gravity methods; Guinea; Guinea Bissau; Ivory Coast; Liberia; Mali; Sierra Leone; surveys; West Africa; West African Shield; applied geophysics. References: 8; illustrations, ISSN: 0764-4450.

Bortnikov, A.Ya. and Barri, M.D. 1995. "Kriterii boksitonosnosti lateritnykh kor vyvetrivaniya Gvinei." Translated "Criteria for bauxite-bearing lateritic weathering crust in Guinea." Izvestiya Vysshikh Uchebnykh Zavedeniy. Geologiya i Razvedka. 1995; 1, Pages 122-125. 1995. Ministerstvo Vysshego i Srednego Obrazovaniya. Moscow, Russian Federation. 1995. Language: Russian. Descriptors: Africa; bauxite; chemically-precipitated-rocks; Devonian; Guinea; laterites;

Liberia; Liberian-Shield; mineral deposits; genesis; Paleozoic; Precambrian; Proterozoic; sedimentary rocks; Silurian; soils; structural controls; upper Precambrian; weathering crust; West Africa; economic geology; geology of ore deposits. References: 3; illustrations, including geological sketch maps. ISSN: 0016-7762.

Brock, M. R.; Chidester, A. H. and Baker, M. W. G. 1977. "Geologic map of the Harper Quadrangle, Liberia. Folio of the Harper quadrangle, Liberia." Miscellaneous Investigations Series Map. Report number: I-0780-D. U. S. Geological Survey, Reston, VA, United States. January 1, 1977. Map Type: geologic map; color map; 44 x 71 cm. on sheet 74 x 106 cm. folded in envelope 30 x 24 cm. Scale 1:250,000 (1 inch = about 4 miles). Sheet 29 by 42 inches. Descriptors: Africa; areal geology; geologic maps; Harper Quadrangle; Liberia; maps; USGS; West Africa. Notes: "Hotines rectified skew orthomorphic projection and rectangular coordinates." "Prepared by the U.S. Geological Survey and the Liberian Geological Survey under the joint sponsorship of the Government of Liberia and the Agency for International Development, U.S. Dept. of State." Includes bibliography. ISSN: 0160-0753; OCLC: 3583320.

Brock, M. R.; Chidester, Alfred H. and Baker, M. W. 1974. "Geology of the Harper Quadrangle, Liberia." Open-File Report. Report number: OF 74-0310. January 1, 1974. Pages (monograph): 12 (1 sheet). U. S. Geological Survey, Reston, VA, United States. Annotation: 1 sheet, scale 1:250,000 (1 inch = about 4 miles). Descriptors: Africa; areal geology; Harper Quadrangle; Liberia; maps; USGS; West Africa. Notes: Map in pocket. Prepared under the auspices of the Govt. of Liberia and the Agency for International Development, U.S. Dept. of State. Transmittal sheet dated November 7, 1974. Bibliography: leaf 12. ISSN: 0196-1497.

Broderick, C. E. 1995. "Changes in the climate at Harbel, Liberia." Biological Agriculture & Horticulture 12, no. 2 (1995) pages 133-149. Abstract: Meteorological data had been routinely collected since 1936 at the Firestone Botanical Research Center, and 53 years of accumulated data were available. The data showed a trend of annual rainfall decline. The number of days with rain per year also declined, and the average temperature at Harbel rose some 0.72 Centigrade degrees over the 53 year period. Number of hours of sunlight per day, as measured by the Stokes sunshine recorder, also increased over the years. The data were found to be accurate, and the manifested trends were very clear. Natural cyclic climatic changes were noted, but deforestation and other events affecting climatic changes were also cited in efforts to explain the noted changes. Descriptors: geographical abstracts; physical geography; Information systems, climatic and soil conditions; Liberia; meteorological data (1936-1989).

Bromery, Randolph Wilson, 1926- . 1968. "Feasibility study for an airborne geophysical survey of the Republic of Liberia." Washington: U.S. Geological Survey, 1968. 23 leaves; 27 cm. Series: Open-file report (Geological Survey (U.S.)); 1137. Notes: Referred to in press release dated December 2, 1968. Bibliography: leaves 21-23. Descriptors: Prospecting; Geophysical methods; Liberia. USGS Library.

Bronevoy, V. A.; Kim, Yu. I. and Kulikova, G. V. 1971. "Osobennosti mineraloobrazovaniya pri formirovanii boksitov v zapadnoy chasti Liberiyskogo shchita." Translated title: "Characteristics of minerals accompanying the formation of bauxites in the western part of the Liberian Shield." In: Kontinental'nyye pereryvy i kory vyvetrivaniya Sibiri. Trudy Sibirskogo Nauchno-

Issledovatel'skogo Instituta Geologii, Geofiziki i Mineral'nogo Syr'ya. 126; Pages 164-167. 1971. Notes: No. 126. Zapadno-Sibirskoye Kniznoye Izdatel'stvo. Novosibirsk, USSR. Language: Russian. Descriptors: Africa; bauxite; economic geology; Liberia; Liberian Shield; mineral composition; West Africa; Economic geology, general. ISSN: 0583-1822.

Bronevoy, V. A.; Ivanov, V. A.; Kim, Yu. I.; Kulikova, G. V.; Mikhaylov, B. M.; Pokrovskiy,V. V.; Safonova, O. F. and Seliverstov, Yu. P. 1970. "Nekotoryye voprosy formirovaniya i razvitiya lateritnykh pokrovov na Liberiyskom shchite (zapadnaya Afrika)." Translated title: "Formation and development of the laterite crust on the Liberian shield, West Africa." Sovetskaya Geologiya. 9; Pages 3-18. 1970. Notes: No. 9. Izdatel'stvo Nedra. Moscow, USSR. Language: Russian. Descriptors: Africa; bauxite; economic geology; geochemistry; laterites; Liberia; Liberian Shield; mineralogy; petrography; soils; West Africa. ISSN: 0038-5069.

Brown, George William. 1941. The economic history of Liberia, by George W. Brown . Washington, D.C., The Associated Publishers, Inc. Description: 366 pages, 24 cm. Notes: Presented as the author's thesis (Ph. D.) at the London School of Economics and Political Science, 1938. cf. Acknowledgements. Includes index. "Selected bibliography": p. 310-323. Subjects: Liberia; Economic conditions. Notes: "The first part is a rapid history and geographic survey, the second part a careful study of the economy of a native communal society, and the "loans and concessions" imposed upon it by the Republic and foreign influences. The appendix quotes historical and economic documents." University of Wisconsin- American Geographical Society Collection. Call Number: HC591.L6 B74 1941.

Brown, Sandra and Gaston, Greg. 2003. "Tropical Africa: Land Use, Biomass, and Carbon Estimates for 1980 (NDP-055)." Note: This document describes the contents of a digital database containing maximum potential aboveground biomass, land use, and estimated biomass and carbon data for 1980. The biomass data and carbon estimates are associated with woody vegetation in Tropical Africa. These data were collected to reduce the uncertainty associated with estimating historical releases of carbon from land use change. Tropical Africa is defined here as encompassing 22.7 x 10{sup 6} km{sup 2} of the earth's land surface and is comprised of countries that are located in tropical Africa (Angola, Botswana, Burundi, Cameroon, Cape Verde, Central African Republic, Chad, Congo, Benin, Equatorial Guinea, Ethiopia, Djibouti, Gabon, Gambia, Ghana, Guinea, Ivory Coast, Kenya, Liberia, Madagascar, Malawi, Mali, Mauritania, Mozambique, Namibia, Niger, Nigeria, Guinea-Bissau, Zimbabwe (Rhodesia), Rwanda, Senegal, Sierra Leone, Somalia, Sudan, Tanzania, Togo, Uganda, Burkina Faso (Upper Volta), Zaire, and Zambia). The database was developed using the GRID module in the ARC/INFO geographic information system. Source data were obtained from the Food and Agriculture Organization (FAO), the U.S. National Geophysical Data Center, and a limited number of biomass-carbon density case studies. These data were used to derive the maximum potential and actual (ca. 1980) aboveground biomass values at regional and country levels. The land-use data provided were derived from a vegetation map originally produced for the FAO by the International Institute of Vegetation Mapping, Toulouse, France. Subjects: Biomass Fuels; Environmental Sciences; Africa; Tropical Regions; Biomass; Carbon; Information Systems; Land Use; Trees; Shrubs. URL: http://www.ornl.gov/~webworks/cppr/y2001/rpt/113771.pdf

Brown Engineers of Liberia, Inc. 1961. Distribution of population, Monrovia area. Brown Engineers of Liberia, Inc. Monrovia: Brown Engineers. Description: 1 map: photocopy; 33 x 55 cm. Cartographic Material. Scale [ca. 1:7,200]. Notes: Shows population by district. At head of Report on water supply, Monrovia, Liberia, 1961. Includes indexed statistical table. "Plate no. 7." Descriptors: Monrovia; Liberia-Population-Maps. LCCN: 96-687121.

Brown Engineers of Liberia, Inc. 1961. Transmission line: [Monrovia region, Liberia]. Brown Engineers of Liberia, Inc. [Monrovia]: Brown Engineers. Description: 1 map: photocopy ; 43 x 44 cm. Cartographic Material. Scale 1:60,000. 1?? = 5,000?. Notes: Shows water lines from White Plains to Monrovia. "April 1961." At head of Report on water supply, Monrovia, Liberia. "Plate no. 3." Descriptors: Water supply-Liberia; Monrovia Region- Maps; Liberia. LCCN: 96-687103.

Brown Engineers of Liberia, Inc. 1961. Water distribution system, Monrovia area. Brown Engineers of Liberia, Inc. Monrovia: Brown Engineers. Description: 1 map: photocopy; 33 x 55 cm. Cartographic Material. Scale [ca. 1:7,200]. Notes: Copy annotated in blue and red crayon to show paved and laterite roads. At head of Report on water supply, Monrovia, Liberia, 1961. "Plate no. 4." Descriptors: Water supply-Liberia; Monrovia- Maps; Monrovia; Liberia-Maps. LCCN: 96-687106.

Burchfield, S. A. 1986. Improving Energy Data Collection and Analysis in Developing Countries: A Comparative Study in Uganda, Liberia and Sudan. Agency for International Development, Washington, DC. Report number: AIDPNAAV672 . June 1986. 355 pages. Descriptors: Information; Government agencies; Developing countries; Uganda; Liberia; Sudan; Energy management; Data collection; Developing country application; Energy analysis; Energy demand; Energy-supplies. Abstract: The study assesses the resources available for collecting/analyzing data in energy planning agencies and organizations in Uganda, Liberia, and Sudan. It examines the quality of the national energy assessments and energy supply/demand balance statements conducted, and makes recommendations on training needs, energy planning activities, and data collection/analysis problems. Interviews were conducted with host government and A.I.D. personnel involved in energy planning activities and projects. Data quality was analyzed using a standardized rating sheet based on recommendations of the U.N. Statistical Commission. The findings identified a number of analytic and institutional problems common to all three countries, and delineated criteria which lead to the success or failure of energy planning activities. NTIS Number: PB87210613XSP.

Burke, Kevin. 1993. "Origin of the rift under the Amazon Basin as a result of continental collision during Pan-African time." In: Geological Society of America, 1993 annual meeting. Abstracts with Programs - Geological Society of America. 25; 6, Pages 233. 1993. Geological Society of America (GSA). Boulder, CO, United States. 1993. Conference: Geological Society of America, 1993 annual meeting. Boston, MA, United States. Oct. 25-28, 1993. Descriptors: Africa; Amazon Basin; Andes; Asia; Atlantic Ocean; Baikal rift zone; Brazil; China; Commonwealth of Independent States; Europe; evolution; Far East; Ghana; Gondwana; gravity anomalies; Keweenawan-Rift; Liberia; North America; Ordovician; Paleozoic; Pan-African-Orogeny; Pangaea; plate collision; plate tectonics; Precambrian; Proterozoic; Rhine Basin; rift zones; Russian Federation; sedimentary rocks; Shanxi-China; Sierra Leone; Silurian; South America; South Atlantic; suture-zones; thermal history; upper Precambrian; West Africa. ISSN: 0016-7592.

Butler, George P. 1978. Realizing the development opportunity [sic] created by an iron ore mining concession in Liberia: the Yekepa model. Corp Partnership for Productivity Foundation, Liberia. Monrovia, Liberia: Sebanoh Printing Press. Description: 109 p., [1] leaf of plates: ill.; 30 cm. Descriptors: Community development-Liberia; Liberia; Yekepa, Liberia-Economic conditions. Named Corp: LAMCO. Notes: Prepared by Partnership for Productivity Foundation/Liberia. LCCN: 79-122682; OCLC: 6917120.

Camil, J.; Tempier, P. and Pin, C. 1983. "Age Liberien des quartzites a magnetite de la region de Man (Cote d'Ivoire) et leur place dans l'Orogene Liberien." Translated title: "Liberian age of magnetite-bearing quartzites from the Man region, Ivory Coast; their place in the Liberian Orogeny." Comptes Rendus des Seances de l'Academie des Sciences, Serie 2: Mecanique-Physique, Chimie, Sciences de l'Univers, Sciences de la Terre. 296; 2, Pages 149-151. 1983. Gauthier-Villars. Montrouge, France. Language: French. Summary Language: English. Descriptors: absolute age; Africa; Archean; dates; evolution; geochronology; isochrons; Ivory Coast; Liberia; Liberian Orogeny; magnetite; Man region; metamorphic rocks; orogeny; oxides; petrology; Precambrian; quartzites; stratigraphy; U-Pb; West Africa. References: 5; illustrations, including table. ISSN: 0750-7623.

Canadian Aero Service, Ltd. 1970. "Preliminary Reconnaissance Survey, Proposed Railway Alignment, Bomi Hills to [the] Mano River." Ottawa, Canada: Canadian Aero Service. 1 map, photocopy, 56 x 192 cm on 2 sheets 62 x 123 cm. and 62 x 136 cm. Scale 1:40,000. Notes: Prepared for mine Management Associates, Ltd., by Canadian Aero Service, Ltd. Oriented with north toward upper left. Relief shown by line forms. "Preliminary information only, subject to revision after completion of a detailed survey. LCCN: gm71-1978.

Carder, Kendall L.; Betzer, Peter R. and Eggimann, Donald W. 1974. Intercomparisons And Application To The West African Shelf. In: Suspended Solids In Water; Marine Science, Volume 4, Proceedings Of Symposium On Suspended Solids In Water, Santa Barbara, California, March 20-22, 1973. Plenum Press, New York, New York, P 173-193, 1974. 7 Fig, 2 Tab, 37 References. ONR N00014-72-0363-0001. Abstract: A property of oceanic particulate matter referred to as 'apparent density' was calculated by dividing the weight of suspended particular matter (spm) by the volume of particles. This parameter is equal to 'mass density' for particles, such as minerals, containing little water. Apparent density calculations were made for a series of samples collected on r/v trident cruise 112 to the continental shelves of Sierra Leone and Liberia. These values ranged from 0.104 to 1.79 for samples with particulate organic carbon fractions (poc/spm) ranging from 0.486 to 0.037. Cross sections of salinity, light scattering beta (45), suspended particulate matter (spm), and beta (45)/total surface area for this region of the west African shelf showed a northwestward-flowing bottom current laden with inorganic sediment having a high apparent density and s southeastward-flowing, organic-rich (low apparent density) surface current. Of the measures of particle concentration applied to these waters, spm and beta (45) showed greatest correlation (r=0.960), suggesting that apparent density is highly correlated with the particle index of refraction. Total particulate volume and total particulate surface area data were not nearly as well correlated with either beta (45) or spm; optical/physical theories were proposed to explain this phenomenon. Descriptors: suspended solids; Continental Shelf; Africa; particle size; Atlantic Ocean; properties; sea water; oceans; on site investigations; sampling; salinity; measurement;

Instrumentation; laboratory tests; physical properties; density; refractivity; analysis; surveys; West African Shelf; Sierra Leone; Liberia; light scattering; Nepheloid Layers.

Carder, Kendall L.; Betzer, Peter R. and Eggimann, Donald W. 1974. "Physical, chemical, and optical measures of suspended-particle concentrations; their intercomparison and application to the West African Shelf." Marine Science (Plenum). 4; Suspended solids in water, Pages 173-193. 1974. Plenum. New York, NY. Descriptors: Africa; bottom currents; concentrations; continental shelf; currents; density; inorganic; Liberia; light scattering; marine transport; measurement; oceanography; organic; RV Trident cruise 112; refractive index; salinity; samples; sedimentation; sediments; Sierra Leone; surface currents; suspended; transport; west; West Africa. Illustrations, including sketch map. Abstract: A property of oceanic particulate matter referred to as 'apparent density' was calculated by dividing the weight of suspended particular matter (spm) by the volume of particles. This parameter is equal to 'mass density' for particles, such as minerals, containing little water. Apparent density calculations were made for a series of samples collected on r/v trident cruise 112 to the continental shelves of Sierra Leone and Liberia. These values ranged from 0.104 to 1.79 for samples with particulate organic carbon fractions (poc/spm) ranging from 0.486 to 0.037. Cross sections of salinity, light scattering beta (45), suspended particulate matter (spm), and beta (45)/total surface area for this region of the west African shelf showed a northwestward-flowing bottom current laden with inorganic sediment having a high apparent density and s southeastward-flowing, organic-rich (low apparent density) surface current. Of the measures of particle concentration applied to these waters, spm and beta (45) showed greatest correlation (r=0.960), suggesting that apparent density is highly correlated with the particle index of refraction. Total particulate volume and total particulate surface area data were not nearly as well correlated with either beta (45) or spm; optical-physical theories were proposed to explain this phenomenon. ISSN: 0160-273X.

Carder, Kendall L; Betzer, Peter R. and McClelland, Scott I. 1974. "Suspended Particle Size Distributions Along the Sierra Leone-Liberian Shelf." Eos, Transactions, American Geophysical Union. 55; 4, Pages 279-280. 1974. American Geophysical Union. Washington, DC, United States. Descriptors: Africa; carbon; continental shelf; detritus; Liberia; models; oceanography; organic carbon; organic compounds; organic materials; phytoplankton; plankton; sea water; Sierra Leone; suspended materials; textures; West Africa; Oceanography. ISSN: 0096-3941.

Carruth, P. 1973. "African Attitudes Toward the Law of the Sea." In: Sea Grant Publication Unc-Sg-73-01, P 166-177, March 1973. 35 References. Abstract: Nations along the gold coast of Africa, such as Ghana, have taken extra measures to protect the wealth that might lie beneath the waters on the extended continental shelf. The Ghana fishery conservation zone presently extends twelve miles off the coast and the territorial continental shelf is restricted to a hundred miles. However, a 1968 law established the territorial seabed at 100 fathoms with an extension to a depth capable of exploitation. Control of coastal oil reserves is one of the possibilities contemplated in this action. The ivory coast has analogous legislation. Kenya, Tanzania, and Liberia all have laws similar to other coastal states, those retaining the twelve mile limit. The Nigerians changed their limit in 1971 from twelve miles to thirty. Massive aid for development and offshore oil deposits are said to have prompted this action. The Nigerian government thinks that it will be able to enforce this boundary and that the country is able to develop this resources in some way economically beneficial to Nigeria. Sea laws that deal with other than local territorial issues are virtually nonexistent. Descriptors: continental shelf; water resources; law of the sea; boundaries, property; *water resources development; oil industry; water law; legislation; political aspects; economic aspects; minerals; governments; International Law; conservation; fishing; social-aspects; legal aspects; natural resources; International Agreements.

Chidester, Alfred Herman. 1972. Project completion report: geological exploration and resources appraisal. Other Titles: Geological exploration and resources appraisal. United States; Agency for International Development; Geological Survey (U.S.). [Washington?]: U.S. Dept. of the Interior, Geological Survey. 34 pages; 27 x 21 cm. Notes: On cover: Project report: Liberia investigations (IR) LI-80. Descriptors: Mines and mineral resources- Liberia; Geology-Liberia; mines and mineral resources; Liberia- Bibliography; Geology- Liberia- Bibliography. OCLC: 1631211.

Chiesa, Sergio. 1991. "El flujo de pomez biotitica del Rio Liberia (Guanacaste), Costa Rica, America Central." Translated "The Rio Liberia biotitic pyroclastic flow (Guanacaste) Costa Rica, Central America." Revista Geologica de America Central. 13; Pages 73-84. 1991. Universidad de Costa Rica, Escuela Centroamericana de Geologia. San Jose, Costa Rica. 1991. Language: Spanish; Summary Language: English. Descriptors: Central America; chemical composition; Costa Rica; crystallization; genesis; Guanacaste; igneous rocks; petrology; pyroclastics; Rio, Liberia; volcanic rocks; Igneous and metamorphic petrology. Illustrations: References: 15; illustrations, including 46 anals., geological sketch map. ISSN: 0256-7024

Choubert, B. 1969. Les Guyano-eburnedes de l'Amerique du sud et de l'Afrique occidentale; essais de comparaison geologique. Translated title: "The geology of Guyana and West Africa; a comparative study." Bulletin du Bureau de Recherches Geologiques et Minieres. Section 4: Geologie Generale. 4; Pages 39-70. 1969. Notes: No. 4. Societe Geologique de France et la Bureau de Recherches Geologiques et Minieres. Paris, France. Language: French. Summary Language: English. Descriptors: Africa; areal geology; correlation; geosynclines; Ghana; Guinea; Guyana; Ivory Coast; Liberia; metals; mineral deposits, genesis; regional; Sierra Leone; South America; West Africa. MAP SCALE: 1:6,000,000. Illustrations, color geologic maps. ISSN: 0153-8446.

Choubert, G. and Faure, Muret A. 1971. "Bouclier eburneen (ou libero-ivoirien)." Translated title: "The "Eburnian" or Liberian-Ivory Coast Shield." In: Tectonique de l'Afrique--Tectonics of Africa. Earth Science (Paris) = Sciences de la Terre (Paris). 6; Pages 185-200. 1971. UNESCO. Paris, France. Language: French; Summary Language: English. Descriptors: Definition and history of shield; geosynclines; granitization, absolute age; Africa; areal geology; coastal; composition; dates; evolution; geochronology; igneous activity; igneous rocks; Ivory Coast; Liberia; metamorphism; petrology; Precambrian; processes; Rb-Sr; structural geology; structure; tectonics; west; West Africa. Illustrations, including geol. sketch map. ISSN: 0070-7910.

Classification and management of soils- Liberia: a four year project design for technical assistance, agricultural sector, Liberia. 1976. MidAmerica International Agricultural Consortium. Design Team. No place of publication given: The Consortium. Description: 65 leaves; 29 cm. Descriptors: Soils- Liberia; Soils- Classification; Liberia. Notes: Contract no. AID/afr-C-1139 (University of Missouri. Bibliography: leaves 64-65. OCLC: 4345802.

Clodius, R.W. 1970. "National Iron Ore Company, Limited; the past, present and future." Bulletin of the Geological, Mining and Metallurgical Society of Liberia. 4; Pages 107-110. 1970. Notes: Vol. 4. Geological, Mining and Metallurgical Society. Monrovia, Liberia. Descriptors: Africa; economic geology; history; iron; Liberia; metals; mining geology; National Iron Ore Company; West Africa. ISSN: 0367-4819.

Coale, William D. 1978. West German trans-nationals in tropical Africa: the case of Liberia and the Bong Mining Company. München: Weltforum-Verlag. 284 pages: ill.; 21 cm. Language: English. Ifo-Forschungsberichte der Abteilung Entwicklungsländer/Afrikastudienstelle; 59; Variation: Ifo-Institut für Wirtschaftsforschung.; Abteilung Entwicklungsforschungsländer-Afrikastudienstelle; Ifo-Forschungsberichte; 59. Descriptors: Corporations; West German- Liberia-Case studies; Corporations, West German- Africa, Sub-Saharan- Case studies. Named Corp: Bong Mining Company. Notes: Originally presented as the author's thesis, Boston University (see below). Bibliography: p. 273-282. ISBN: 3803901650; LCCN: 79-343884; OCLC: 4774640.

Coale, William D. 1977. West German trans-nationals in tropical Africa: the case of Liberia and the Bong Mining Company. Description: xviii, 285 leaves; 29 cm. Descriptors: International business enterprises; Mineral industries- Liberia; Liberia- Relations- Germany; Liberia. Notes: Photocopy of typescript. Vita. Includes bibliographical references (leaves 274-282). Dissertation: Thesis, Boston University. OCLC: 3215317.

Coleman, Ciyata Dinah. 1985. Exports and spatial economic growth in Liberia: effects of rubber and iron ore. Thesis (Ph. D.) Southern Illinois University at Carbondale, 1986. Microfilm. Ann Arbor, Mich: University Microfilms International. 1 microfilm reel; 35 mm. Description: 95 leaves: ill., map; 29 cm. Descriptors: Export marketing- Liberia; Export sales contracts- Liberia; Rubber industry and trade- Liberia; Iron mines and mining- Liberia; Liberia- Economic conditions. Notes: Vita. Includes bibliographical references (leaves [92]-97). OCLC: 15210312.

Commission for Technical Co-Operation in Africa South of the Sahara. Commission for Technical Co-Operation in Africa. "Geological bibliography of Africa south of the Sahara = Bibliographie géologique de l'Afrique au sud du Sahara. Commission for Technical Co-Operation in Africa South of the Sahara.; Inter-African Scientific Correspondent for Geology. ; Commission for Technical Co-Operation in Africa.; Inter-African Scientific Correspondent for Geology. London: Commission for Technical Co-operation in Africa South of the Sahara, Inter-African Scientific Correspondent for Geology. 1950s-? Language: English; In English and French. Descriptor: Geology- Africa, Sub-Saharan- Bibliography- Periodicals. Note(s): Each no. has also a distinctive title. [Vol.] 1-2 issued by the Inter-African Scientific Correspondent for Geology, Commission for Technical Co-operation in Africa South of the Sahara; [v.] 3- issued by the Inter-African Scientific Correspondent for Geology, Commission for Technical Co-operation in Africa. OCLC: 8007183.

Conover, Helen Field. 1957. "Liberia." In: Africa South of the Sahara: A Selected and Annotated List of Writings, 1951-1956. Compiled by Helen Conover. Library of Congress, Reference Department, General Reference and Bibliography Division. The section on Liberia is between pages 248-255. OCLC: 01869067.

Conover, Helen Field. 1952. "Liberia." In: Introduction to Africa; a selective guide to background reading. Library of Congress. European Affairs Division. Washington, DC: University Press of Washington. 237 pages, map. 25 cm. Notes: "Liberia" is between pages 224 and 228. ISBN: 0837118379; LCCN: 52-60007; OCLC: 553282

Coonrad, W. L. 1979. "Mineral map of the Bopolu Quadrangle, Liberia." Liberia.; United States. Agency for International Development. Open-File Report. Report number: OF 79-1516. Pages: 27. Project report. Liberian investigations; (IR)LI-61[E;] Open-file report (Geological Survey (U.S.)); 79-1516. U. S. Geological Survey, Reston, VA, United States. January 1, 1979. 25 pages: 1 color map (in pocket); 27 cm. 1 over-size sheet, scale 1:250,000 (1 inch = about 4 miles). Descriptors: Africa; Bopolu Quadrangle; economic geology; economic geology maps; explanatory text; Liberia; maps; mineral resources; USGS; West Africa. Notes: Prepared under the auspices of the Government of Liberia and the Agency for International Development, U.S. Department of State. Includes bibliographical references (p. 12-14). ISSN: 0196-1497; OCLC: 7029214. Online: http://onlinepubs.er.usgs.gov/djvu/OFR/1979/ofr_79_1516.djvu

Coonrad, W. L.; Cooper, B. R. and Phillips, E. 1979. "Mineral Localities of the Zwedru Quadrangle, Liberia, 1:250,000 scale map." US Geological Survey, Liberia Geological Survey and US Agency for International Development, Project Report (IR) LI 70-E. AMRS, Inc. See: http://www.africaminerals.com/

Coonrad, W. L.; Dunbar, J. D. N. and Dinkins, S. 1979. "Mineral Localities of the Gbanka Quadrangle, Liberia, 1:250,000 scale map." US Geological Survey, Liberia Geological Survey and US Agency for International Development, Project Report (IR) LI 63-E. AMRS, Inc. See: http://www.africaminerals.com/

Coonrad, W. L.; Phillips, E.; Sherman, T.W. and Shannon, E.K. 1979. "Mineral Localities of the Juazohn Quadrangle, Liberia, 1:250,000 scale map." US Geological Survey, Liberia Geological Survey and US Agency for International Development, Project Report (IR) LI 65-E. AMRS, Inc. see: http://www.africaminerals.com/

Coonrad, W. L.; White, R.W. and Stewart, W. 1979. "Mineral Localities of the Voinjama Quadrangle, Liberia, 1:250,000 scale map." US Geological Survey, Liberia Geological Survey and US Agency for International Development, Project Report (IR) LI 68-E. AMRS, Inc. See: http://www.africaminerals.com/

Cooper, B. R. 1963. "A Geologic Report of Sections of Zorzor and Voinjama Districts, Western Province." An unpublished un-numbered Geological Survey of Liberia Report, 11 pages. AMRS, Inc. See: http://www.africaminerals.com/

Cora, Guido. 1892? "Notizie sulla repubblica di Liberia specialmente secondo I viaggi e gli studo di J. Büttikofer…" Translated "News on the Republic of Liberia especially the second expedition and the studies of J. Büttikofer." Publication: Torino, 1892? Series: Estratto dal Cosmos di Guido Cora, Serie II, volume XI (1891-1892). Held at the USGS Library, Pamphlet Collection.

Cortesini, Augusto and Minner, John R. 1973. "Petroleum Developments in Central and Southern Africa in 1972." The American Association of Petroleum Geologists Bulletin. 57; 10, Pages 2008-2056. 1973. American Association of Petroleum Geologists. Tulsa, OK, United States. Descriptors: 1972; Africa; Angola; Benin; central; Central Africa; Congo; East Africa; economic geology; Ethiopia; Gabon; Gambia; gas; Ghana; Guinea; Indian Ocean Islands; Ivory Coast; Kenya; Lesotho; Liberia; Madagascar; Mali; Mauritania; Mauritius; Mozambique; natural gas; Nigeria; petroleum; petroleum exploration; production; south; South Africa; Southern Africa; West Africa; sketch maps. ISSN: 0002-7464.

Coury, A.B. and Hendricks, T.A. 1978. "Map of Prospective Hydrocarbon Provinces of the World, Europe, West Asia, Africa." US Geological Survey Map MF-1044B, scale 1:20,000,000. USGS Library.

Croze, W. W. J. 1939. Liberia: report on iron ore occurrences in the northwestern part of Liberia, West Africa. United States Steel Corporation. No place of publication given. Description: 105 leaves, [5] leaves of plates: ill.; 28 cm. Descriptors: Geology- Liberia; Iron ores- Liberia; Liberia. Notes: Typescript (carbon copy). Issued as a reconnaissance survey for the United States Steel Corporation. OCLC: 20228976.

Culver, Stephen J. 1988. "The Pan-African Rokelides of Sierra Leone; problems and possible correlations with Liberia and Guinea." In: The West African connection; evolution of the central Atlantic Ocean and its continental margins. Sougy, Jean (editor); Rodgers, John (editor). Journal of African Earth Sciences. 7; 2, Pages 514. 1988. Pergamon. London-New York, International. 1988. Conference: The West African connection; evolution of the central Atlantic Ocean and its continental margins. Giens, France. Jan. 17-22, 1984. Descriptors: Africa; Bania Group; Cambrian; Guinea; Kasila Group; Kerema Assemblage; Kora Group; Liberia; Marampa Group; Mount Gibi Formation; Paleozoic; Pan African Orogeny; Precambrian; Proterozoic; Rokel River Group; Rokelides; Sangoya-Group; Sierra Leone; stratigraphy; upper Precambrian; West Africa. ISSN: 0731-7247

Culver, Stephen J. and Magee, A. W. 1987. "Late Precambrian glacial deposits from Liberia, Sierra Leone, and Senegal, West Africa." National Geographic Research. 3; 1, Pages 69-81. 1987. National Geographic Society. Washington, DC, United States. Descriptors: Africa; ancient ice ages; clastic rocks; environment; glacial environment; glacial extent; glacial geology; Liberia; Precambrian; sedimentary rocks; sedimentation; Senegal; Sierra Leone; stratigraphy; tillite; West Africa. References: 28; illustrations, including sketch maps. ISSN: 8755-724X.

Culver, S. J; Williams, H. R. and Venkatakrishnan, R. 1991. "The Rokelide Orogen." In: The West African orogens and circum-Atlantic correlatives. Dallmeyer, R. D. (editor). Pages 123-150. 1991. Springer-Verlag. Berlin, Federal Republic of Germany. 1991. IGCP (International Geological Correlation Programme). Conference: Tectonothermal evolution of the West African orogens and circum-Atlantic terrane linkages. Nouakchott, Mauritania. Dec. 1987. Descriptors: Africa; anomalies; foliation; geophysical methods; geophysical surveys; gravity methods; Guinea; IGCP; Liberia; magnetic methods; metamorphism; orogenic belts; Paleozoic; radioactivity methods; Rokelide-Orogeny; seismic methods; Sierra Leone; stratigraphy; structural geology; surveys; tectonics; tectonostratigraphic units; West Africa; applied geophysics. Notes: IGCP Project No.

233. Illustrations: References: 85; illustrations, including sects., strat. color, geol. sketch maps. ISBN: 3-540-52412-6; 0-387-52412-6.

Dallmeyer, R. D. 1989. "Contrasting accreted terranes in the Southern Appalachian orogen and Atlantic-Gulf Coastal plains and their correlations with West African sequences." In: Terranes in the Circum-Atlantic Paleozoic orogens. Special Paper - Geological Society of America. 230; Pages 247-267. 1989. Geological Society of America (GSA). Boulder, CO, United States. 1989. IGCP (International Geological Correlation Programme). Descriptors: accretion; Africa; Appalachian Phase; Appalachians; Ar-Ar; Atlantic Coastal Plain; basement; Bassaride Orogeny; correlation; crystallization; decollement; evolution; faults; Florida; Guinea; Gulf Coastal Plain; IGCP; intrusions; kinematics; lateral faults; Liberia; Mauritanide Orogeny; metamorphic rocks; North America; orogeny; Paleozoic; Pangaea; Permian; Piedmont; plate collision; plutons; Rb-Sr; right lateral faults; Rokelide Orogeny; Senegal; Sierra Leone; Southern Appalachians; structural geology; terranes; transcurrent faults; United States; West Africa; Wiggins Arch. Notes: IGCP Project No. 233. Illustrations: References: 120; illustrations, including sections, geological sketch maps. ISSN: 0072-1077; ISBN: 0-8137-2230-6.

Dalrymple, G. B.; Grommé, C. Sherman and White, Richard W. 1975. "Potassium-argon age and paleomagnetism of diabase dikes in Liberia; initiation of central Atlantic rifting." Geological Society of America Bulletin. 86; 3, Pages 399-411. 1975. Geological Society of America (GSA). Boulder, CO. Abstract: Tholeiitic diabase dikes that trend northwest-southeast, parallel to the coastline, are common in northwestern Liberia. K-Ar whole-rock and mineral ages determined from dikes that intrude Precambrian crystalline rocks are discordant and range from 186 to 1,213 m.y. Incremental heating experiments on three neutron-irradiated samples of these rocks give "saddle-shaped" $^{40}Ar/^{39}Ar$ release diagrams that reach minima of less than 300 m.y. at intermediate temperatures and that do not fit a $^{40}Ar/^{36}Ar$ versus $^{39}Ar/^{36}Ar$ isochron. K-Ar ages determined from diabase dikes and sills that intrude Paleozoic sedimentary rocks near the coast are all within the range 173 to 192 m.y. $^{40}Ar/^{39}Ar$ incremental heating data for one of these samples gives a plateau age and a $^{40}Ar/^{36}Ar$ versus $^{39}Ar/^{36}Ar$ isochron age that are concordant with the conventional K-Ar age. The conventional and $^{40}Ar/^{39}Ar$ K-Ar data show that the dikes intruding Precambrian basement rocks contain large and variable amounts of excess ^{40}Ar, whereas the diabase intruding Paleozoic sandstone does not. All of the intrusions are probably earliest Jurassic in age. Mean paleomagnetic directions in six dikes and sills that intrude sedimentary rocks are nearly parallel to mean paleomagnetic directions in 19 dikes that intrude Precambrian rock, further evidence for contemporaneity. The paleomagnetic pole derived from all 25 diabase units is at lat 68° N., long 242° E., with $\alpha_{95} = 5°$, in close agreement with other Mesozoic paleomagnetic poles from the African continent. A mean paleomagnetic pole for northwest Africa has been calculated using these data and published paleomagnetic directions from 19 other intrusive rock units that have similar radiometric ages in Morocco and Sierra Leone. This pole is compared with another paleomagnetic pole calculated from published data from 16 localities in igneous rocks of latest Triassic to earliest Jurassic age distributed from Nova Scotia to Pennsylvania. The comparison shows that, with the African and North American continents in their present positions, the two poles differ by 44° of arc, but when the continents are restored to the predrift configuration proposed by Bullard and others (1965), the angular difference diminishes to 3°. This coincidence of paleomagnetic poles provides an earliest limit of 180 ± 10 m.y. for the separation of Africa from North America. Descriptors: absolute age; Africa; age; Ar-Ar; Atlantic Ocean; continental drift; data; dates;

diabase; dikes; geochemistry; igneous rocks; intrusions; K-Ar; Liberia; Mesozoic; minerals; North America; northwest; paleomagnetism; Phanerozoic; plutonic rocks; pole positions; reconstruction; rifting; sea floor spreading; South America; tectonophysics; tholeiitic; West Africa; whole rock. ISSN: 0016-7606.

Damme, Wim van. 1995. "Do refugees belong in camps? Experiences from Goma and Guinea. (Viewpoint)." The Lancet. Volume 346 (8971). August 5, 1995: pages 360 et seq. Abstract: There are alternatives to African refugee camps and their associated health problems. In 1994, refugees from Rwanda arrived in large numbers in Zaire and were subsequently marched to three camps outside of Goma. Without water, proper latrines and burial grounds, cholera broke out. Refugees have been dependent on outside help without adequate food supplies. Alternatively, in 1989, refugees from Liberia and Sierra-Leone settled in Guinea in border villages and small towns. The government did not establish camps for refugees but subsidized villages that welcomed refugees. Refugees had free health care access to existing medical facilities. Epidemics occurred, but on a much smaller scale comparable to the local rate of disease. Furthermore, this population increase may have created an economic boom for Guinea.

Daves, L. C. and Gill, W. H. 1922. Preliminary base map of the Republic of Liberia. Liberian Cartographic Service. and United States Department of State. Washington, D.C.: Dept. of State. Description: 1 map: color; 90 x 78 cm., on sheet 99 x 82 cm. Descriptors: Liberia-Maps. Map Info: Scale 1:600,000. Notes: Relief shown by hachures. Includes inset showing sources of cartographic information. US Department of State; Liberian Surveys; compiled at Monrovia, June 1922 by L.C. Daves; prepared for engraving, Washington, D.C., by W.H. Gill. Planetable reconnaissance controlled by triangulation from Zorzor base- Zigida datum. Exploratory surveys with sketching board- controlled by sextant observations. Military route sketches controlled by frequent compass and times distance with a few sextant observations. British ordnance maps corrected (sheets 59T-Kissi and 71B-Gola. Anglo-Liberian Boundary Commission maps 1903 and 1913-1914. Maps of Commission de Delimtation Franco-Liberianne. Coast from British Admiralty charts. OCLC: 55093231.

Davis, Paul K. 1973. "Effect of Pressure Shock on Argon 40 39 Dating." In: Fall Annual Meeting, San Francisco, 1973; Section of Volcanology, Geochemistry, and Petrology; Experimental and Theoretical Petrology. Eos, Transactions, American Geophysical Union. 54; 11, Pages 1223. 1973. American Geophysical Union. Washington, DC. Descriptors: absolute age; Africa; Apollo 14; Apollo Program; Ar-36; Ar-39; Ar-40; Ar-40-Ar-39; argon; basalts; Brent Crater; Canada; comparison; Eastern Canada; effects; geochronology; igneous rocks; isotopes; Liberia; mass; methods; Moon; Nipissing District, Ontario; noble gases; Ontario; pressure; radioactive isotopes; shock; spectroscopy; stable isotopes; volcanic; volcanic rocks; West Africa; Geochronology. ISSN: 0096-3941.

Delor, C.; Lafon, J. M.; Krymsky, R.; Luais, B.; Milesi, J. P.; Phillips, D. and Rombouts, L. 2004. "Mid-Jurassic ages for West African Kimberlites: First U-Pb, Sm-Nd and Ar-Ar data." 20ème Colloque de Géologie Africaine - 20th Colloquium of African Geology, BRGM, Orléans, France, 2-7 Juin 2004. Orléans, France: BRGM, GSAF, CNRS, COGEMA. Page 129. Descriptors: Geochronology.

Delteil, J. R.; Valery, Pierre; Montadert, Lucien; Fondeur, C.; Patriat, Philippe and Mascle, Jean. 1974. "Continental margin in the northern part of the Gulf of Guinea." In: The geology of continental margins. Burk, Creighton A. and Drake, Charles L, eds. Pages 297-311. 1974. Springer-Verlag. New York, United States. Descriptors: Africa; Atlantic Ocean; bathymetry; Benin; continental margin; continental shelf; geophysical methods; geophysical surveys; Ghana; Guinea; Gulf of Guinea; Ivory Coast; Liberia; marine geology; Niger Delta; Nigeria; North Atlantic; oceanography; profiles; reflection; sediments; seismic methods; structure; surveys; thickness; Togo; Walda; West Africa. Illustrations, including sketch maps.

Desk Study on the Environment in Liberia. 2004. United Nations Environment Programme. 116 pages. Abstract: The United Nations Environment Programme (UNEP) participated in the post-conflict UN Needs assessment mission to Liberia. The findings of this desk study draw on key environmental information obtained from the Liberian national authorities, non-governmental organisations and other sources. Recommendations are made how environment could be fully integrated into the coming reconstruction efforts in Liberia. This study of Liberia shows how environment and development are fully interlinked even in the poorest societies. The clean-up of the environment after the conflict period and sustainable management of natural resources are prerequisites for the safe return of refugees, sound livelihoods and successful reconstruction of the country. ISBN: 9280724037.

Dias, Edelberto S.; Fortes-Dias, Consuelo L.; Stiteler, John M.; Perkins, Peter V. and Lawyer, Phillip G. 1998. "Random amplified polymorphic DNA (RAPD) analysis of Lutzomyia longipalpis laboratory populations." Revista do Instituto de Medicina Tropical de São Paulo. Brazil Jan./Feb. 1998, volume 40 (1), pages 49-54. Note: The phlebotomine sand fly Lutzomyia longipalpis has been incriminated as a vector of American visceral leishmaniasis, caused by Leishmania chagasi. However, some evidence has been accumulated suggesting that it may exist in nature not as a single but as a species complex. Our goal was to compare four laboratory reference populations of L. longipalpis from distinct geographic regions at the molecular level by RAPD-PCR. We screened genomic DNA for polymorphic sites by PCR amplification with decamer single primers of arbitrary nucleotide sequences. One primer distinguished one population (Marajó Island, Pará State, Brazil) from the other three (Lapinha Cave, Minas Gerais State, Brazil; Melgar, Tolima Department, Colombia and Liberia, Guanacaste Province, Costa Rica). The population-specific and the conserved RAPD-PCR amplified fragments were cloned and shown to differ only in number of internal repeats. Subjects: RAPD-PCR; Finger printings; Sand fly; Lutzomyia; Genotyping. ISSN 0036-4665.

DiCarlo, Carmen and Quick, James Robert. 1968. Infrared Mapping of Liberia. US Army Engineer Topographic Labs, Fort Belvoir VA, United States. May 1968. 23 Pages(s). Abstract: Briefly discussed in this paper is the country of Liberia - its land, its economy and its people. Also reviewed, in somewhat greater detail, are the activities associated with Project 67-4 - the mapping of Liberia using infrared film - from January 1967 through the completion of field operations in March 1968. Descriptors: Sub-Saharan Africa; photogrammetry; infrared photography; photogrammetry; natural resources; visual perception; photography; infrared equipment; stabilization; camera mounts; training; Army personnel; weather stations. Prepared in cooperation with Defense Intelligence Agency, Washington, D. C. Approved For Public Release. DTIC: AD0721911.

Dijkstra, K-D. B. and Lempert, J. 2003. "Odonate assemblages of running waters in the Upper Guinean forest." Archiv fuer Hydrobiologie [Arch. Hydrobiol.]. Vol. 157, no. 3, pages 397-412. Jun 2003. Descriptors: community composition; habitat selection; streams; Odonata; Liberia; Ghana. Abstract: In order to describe the assemblages of adult Odonata of running waters in the Upper Guinean forest, 36 sites in Liberia and Ghana were analyzed using Non-metric Multidimensional Scaling. Five groups were identified, which correspond with different assemblages in the sequence of habitats from small streams to large rivers. Taxonomically related species demonstrate distinct ecological segregation within this gradient, occupying different sections of running waters, or different microhabitats therein. The balance of sun and shade, resulting from a varying degree of habitat openness, is thought to be an important factor in habitat selection, but it is difficult to distinguish from other factors associated with stream size. Anthropogenic opening of stream habitat (e.g. by deforestation or damming) can downscale the present fauna, i.e. result in the invasion of species of downstream habitats (more open) and the disappearance of upstream (dense forest) species. ISSN: 0003-9136.

Dillon, William P. and Sougy, Jean M. A. 1974. "Geology of West Africa and Canary and Cape Verde Islands." In: The Ocean Basins and Margins; Vol. 2, The North Atlantic. Pages 315-390. 1974. Plenum Press, New York. Descriptors: Africa; Anti Atlas; areal geology; Atlantic Ocean; Atlantic Ocean Islands; Atlas Mountains; basement; basins; Benin; Canary Islands; Cape Verde Islands; Cenozoic; complexes; continental margin; cratons; crust; evolution; Gambia; geochronology; Ghana; Guinea; Guinea Bissau; Hercynian; igneous rocks; intrusions; islands; Ivory Coast; Kayes Massif; Leo Uplift; Liberia; mantle; Mauritania; Mesozoic; metamorphism; Moroccan Atlas Mountains; Morocco; Nigeria; North Africa; ocean basins; orogeny; outcrops; Paleozoic; Pan African; petrology; Precambrian; Reguibat Massif; sedimentary petrology; sedimentary rocks; sedimentation; Senegal; Sierra Leone; stratigraphy; structural geology; tectonics; Togo; volcanic; volcanism; West Africa; West African-Shield; Western Sahara; zoning; Illustrations, including geological sketch maps.

Dobson, J. E. 1984. "Energy for Liberia: problems and options." 450 pages. Conference: Association of American Geographers Annual Meeting "Geography and Public Policy", Washington, DC (USA), 22-25 April 1984. (World Meeting Number 842 0080). Notes: Abstracts book available: Association of American Geographers, 1710 Sixteenth St., N.W., Washington, DC 20009, USA, ISBN 0-89291-176-X.

Donahue, C. A.; Traxler, K. M.; Walters, K. R.; Louer, J. W. and Gilford, M. T. 1995. Equatorial Africa. A Climatological Study. Technical note. March 1995. 276p. Report number: USAFETACTN95001. Performer: Air Force Environmental Technical Applications Center, Scott AFB, IL. Descriptors: control; clouds; weather; Sub-Saharan Africa; hazards; precipitation; visibility; climate; meteorology; Parks; north direction; geography; south direction; Nigeria; commonality; Angola; Mozambique; seasons; Zaire; Kenya; West Africa; Ghana; Ivory Coast; Liberia; Malawi; Mauritania; Tanzania. Equatorial regions; wind; climatology; Africa. Abstract: A climatological study of Equatorial Africa, a region that comprises Senegal, Gambia, Guinea-Bissau, Guinea, Sierra Leone, Liberia, Ivory Coast, Ghana, Togo, Benin, Nigeria, Cameroon, Equatorial Guinea, Gabon, Congo, Central African Republic, Rwanda, Burundi, Uganda, Tanzania, Kenya, Malawi, the southern parts of Mauritania, Mali, and Niger, and the northern parts of Angola, Mozambique, Zaire, and Zambia. After describing the geography and major meteorological features of the entire region, the study discusses the climatic controls of each of Equatorial Africa's eight 'zones of climatic commonality' in detail. Each, 'season' is defined and discussed in considerable detail, to include general weather, clouds, visibility, winds, precipitation, temperature, and other hazards. DTIC Number: ADA2939569XSP; ProxyURL/Handle: http://handle.dtic.mil/100.2/ADA293956

Donner, Etta, 1911- . 1939. Hinterland Liberia, by Etta Donner. Translated by Winifred M. Deans. London, Glasgow: Blackie & Son, Limited. Description: 302 p. maps, plates. 22 cm. Notes: Map on lining-papers. "First published 1939." Descriptors: ethnology Liberia; Liberia Description and travel; Liberia Social life and customs. Other: Deans, Winifred M. (Winifred Margaret) translator. University of Wisconsin- American Geographical Society Collection. Call Number: DT626 .B36 1939.

Dorbor, Jenkins K. 1998. "Liberia." Mining Annual Review. 1998; Pages 200. 1998. Mining Journal. London, United Kingdom. 1998. Descriptors: Africa; diamonds; gold ores; industrial minerals; iron ores; Liberia; metal ores; mineral resources; mining legislation; reserves; review; West Africa; Economic geology, general, deposits. ISSN: 0076-8995.

Dorbor, Jenkins K. 1987. "The Precambrian of Liberia; some chemical features of Liberian granitoids." Compass of Sigma Gamma Epsilon, 1915-84. 64; 4, Pages 244-263. 1987. National Council of Sigma Gamma Epsilon, University of Oklahoma. Norman, OK, United States. Descriptors: Africa; geochemistry; granites; igneous-rocks; Liberia; petrology; plutonic-rocks; Precambrian; West Africa. References: 27; illustrations, including 2 tables, geol. sketch map. ISSN: 0010-4213.

Dorbor, Jenkins K. and Konuwa, J. B. 1987. "Progress Report on the Henry Town/Tawalata Geological and Mineral Exploration and Quantification Project From December 6-18, 1987." Geological Survey of Liberia unpublished un-numbered report, 4 pages. AMRS, Inc. See: http://www.africaminerals.com/

Dorbor, Jenkins K. and Konuwa, J. B. 1988. "Progress Report on the Henry Town- Tawalata Geological and Mineral Exploration Project January 22 - February 19, 1988." Geological Survey of Liberia unpublished un-numbered report, 3 pages. AMRS, Inc. See: http://www.africaminerals.com/

Dorbor, Jenkins K. and Nair, A. M. 1989. "African geology; Liberian geology; status, progress and needs." Bulletin Africain de Geoscience = African Geoscience Newsletter. 35; Pages 18-21. 1989. Ahmadu Bello University, Department of Physics and Geology, ... for the AGID and GSA (Geological Society of Africa). Zaria, Nigeria. 1989. Descriptors: Africa; areal geology; Bong Range; concepts; Gibi Mountain; gneisses; granite gneiss; Guinean Shield; Liberia; lithostratigraphy; metamorphic rocks; Nimba Mountains; regional; research; West Africa; Wologizi Mountain. ISSN: 0189-9392

Dorbor, Jenkins and Wright, Leslie. 1999. "Liberia." Mining Annual Review. 1999; Pages 88. 1999. Mining Journal. London, United Kingdom. 1999. Descriptors: Africa; legislation; Liberia;

mineral economics; mineral exploration; mineral resources; mining; policy; West Africa; Economic geology; general economics. ISSN: 0076-8995.

Dorm-Adzobu, C. 1985. Forestry and Forest Industries in Liberia. An Example in Ecological Destabilization. Performer: Wissenschaftszentrum, Berlin (Germany, F.R.). Internationales Inst. fuer Umwelt und Gesellschaft. 25pages. Advisory: Also available from Wissenschaftszentrum Berlin (Germany, F.R.). Internationales Inst. fuer Umwelt und Gesellschaft. Descriptors: Wood-products; agriculture; deforestation; economics; risk; revegetation; forests; combustions; area; Liberia; forestry; foreign technology; natural resources and earth sciences forestry. Abstract: The forestry sector in Liberia is the third most important sector of the national economy in terms of foreign exchange after iron ore mining and rubber production. Although the commercial exploitation of Liberia's forests only began in the mid-sixties, the forest cover has already been lost to such a degree that the extension of the Sahel drought into Liberia has become a serious threat. The risk of continuing traditional forest policy has been clearly recognized by the Liberian government and since 1973 the government has been trying to put the use of the forests on a sustainable basis. Furthermore, the slash and burn techniques of forest use by local farmers continues unabated and its impact on deforestation remains uncontrolled. As a result the ratio between reforestation and deforestation reached 1:20 in terms of acreage until 1981, but fell to 1:100 thereafter. NTIS Number: TIBB8706482XSP.

Douglass, William. 1867. "Sketch of the Careysburg Road, etc." Liberia; Careysburg Region. Description: 2 profiles on 1 sheet: manuscript, color; sheet 32 x 49 cm. Series: [American Colonization Society map collection; 20]. Contents: From Careysburg, to Receptacle Hill- From Receptacle Hill, to St. Paul's River. Descriptions: roads- Liberia; Careysburg Region-Charts diagrams. Maps, Manuscript. Notes: Pen-and-ink and pencil. Includes indexed table and notes. Available also through the Library of Congress Web site as a raster image. Map Info: Not drawn to scale. Made by William Douglass, December, 1867. LCCN: 96-684996; OCLC: 37986409; Access Path: http://hdl.loc.gov/loc.gmd/g8884c.lm000006

Drechsel, P.; Schmall, S. and Zech, W. 1990. "Relationships between growth, mineral nutrition, and soils in young teak plantations in Benin and Liberia." Water, Air & Soil Pollution. Vol. 54, no. special, pages 651-656. 1990. Conference: International Symposium on Management of Nutrition in Forests Under Stress, Freiburg (Germany), 18-21 September 1989. Descriptors: trees; growth; minerals; Tectona grandis; Liberia. Abstract: Growth and vigor of trees show considerable variations in young teak plantations in Benin (vertisols) as well as in Liberia (Ferralsols). Differences in growth are mainly related to topsoil acidity and the foliar Ca-status in Liberia. In Benin, water logging (followed by root decay) reduces the Mg super(-), K super(-) and N-uptake. In addition, growth on the Vertisols is limited by a low K sub(EX)/CEC-ratio. ISSN: 0049-6979.

Drechsel, P. and Zech, W. 1994. "DRIS evaluation of teak (Tectona grandis L.f.) mineral nutrition and effects of nutrition and site quality on teak growth in West Africa." Forest Ecology and Management. Vol. 70, no. 1-3, pages 121-133. 1994. Descriptors: nutrition; minerals; growth; soil analysis; Tectona grandis; Africa. Abstract: The objective of the investigation was to study the site variables controlling teak yield (Tectona grandis Linn.fil.) and to establish guidelines for the selection of high productivity sites in Benin, Cote d'Ivoire, Liberia, Nigeria and Togo. Depending on stand age, soil and region, between 70 and 90% of the variation in tree growth (site index, SI) could be explained by the supply of nitrogen, the root-available soil depth and precipitation. Diagnostic foliar analysis for a broad range of elements was carried out in all plantations with the exception of Nigeria. This showed that in 20% of the stands, various deficiency symptoms occur, and in an additional 40%, hidden demand of at least one nutrient is apparent. According to the Diagnosis and Recommendation Integrated System (DRIS), the most deficient nutrients besides N are Ca and P, while in 45% of all stands there is a relative Al excess. Recommendations for the evaluation and classification of site quality and the number of trees sampled for foliar analysis are given. ISSN: 0378-1127.

Dupuy, C.; Marsh, J.; Dostal, J.; Michard, A. and Testa, S. 1988. "Asthenospheric and lithospheric sources for Mesozoic dolerites from Liberia (Africa); trace element and isotopic evidence." Earth and Planetary Science Letters. 87; 1-2, Pages 100-110. 1988. Elsevier. Amsterdam, Netherlands. Descriptors: Africa; alkaline-earth-metals; asthenosphere; classification; diabase; dikes; genesis; geochemistry; igneous-rocks; intrusions; isotopes; Liberia; lithosphere; major elements; mantle; Mesozoic; metals; Nd144- Nd143; neodymium; plate tectonics; plutonic rocks; rare earths; rifting; Sr87, Sr86; stable isotopes; strontium; trace elements; upper mantle; West Africa. References: 40; illustrations, including 24 anal., 3 tables, sketch map. ISSN: 0012-821X.

Durrant, Robert Ernest, 1906- . 1925. Liberia: a report by Robert Ernest Durrant for the African International Corporation (1924) Limited. London: African International Corporation (1924) Limited, [1925]. Description: Book. 71 pages: ill., plates, portraits, folded map; 24 cm. Notes: Edition statement on title page verso appears to be printing information. Descriptors: Liberia. University of Wisconsin- American Geographical Society Collection. Call Number: DT624 .D87x 1925.

Durazzo, A. and Gordon, R. L. 1967. Liberia: Mineral Wealth, Private Capital and Economic Development. Mining Magazine. Volume 117, no 5, November 1967, pages 330-339. "Dutch Exploration in Liberia." 1967. Mining Journal. Number 269, page 69 et seq. ISSN: 0026-5225; OCLC: 02438984.

Eggimann, Donald W.; Betzer, Peter R. and Carder, Kendall L. 1980. "Particle transport from the West African shelves of Liberia and Sierra Leone to the deep sea; a chemical approach." Marine Chemistry. 9; 4, Pages 283-306. 1980. Elsevier: Amsterdam, Netherlands. Descriptors: Africa; Atlantic Ocean; chemical composition; composition; data; deep sea environment; East Atlantic; geochemistry; Liberia; marine environment; marine transport; ocean-circulation; oceanography; provenance; sea water; sedimentation; Sierra Leone; suspended materials; transport; West Africa. Abstract: Suspended particle transport through the sea was examined over a 480-km section of the Liberia Sierra Leone continental shelf off West Africa using the chemical and mineralogical composition of the particles as tracers. Ratios of Si/Al, Fe/Al, Mg/Al, and Mn/Al were used to detect shelf-derived matter in slope waters. The use of light-scattering and particle-mass measurements was not applicable to most of the material analyzed, so that chemical identification of particulates was necessary. Suspended particulates and sediments from the adjacent eastern Atlantic basin were similar in chemical and mineralogical composition to those in the water column seaward of the West African continental shelves. However, these basin specimens were distinct from Sahelian dust, which is thought to be the main source of sedimentary matter for tropical and semi-tropical regions of the deep eastern basin. These results indicate that shelf input of matter to the deep sea may be greater than was previously suspected. References: 39; illustrations, including tables, sketch map. ISSN: 0304-4203.

Eggimann, D. W.; Betzer, P. R. and Carder, K. L. 1975. "A chemical study of particle transport from the West African Shelf to the deep ocean." Eos, Transactions, American Geophysical Union. 56; 6, Pages 371. 1975. American Geophysical Union. Washington, DC, United States. Descriptors: Africa; Atlantic Ocean; chemical composition; clay minerals; continental shelf; currents; kaolinite; Liberia; oceanography; sedimentation; sediments; sheet silicates; Sierra Leone; silicates; southeast; suspended materials; transport; West Africa. ISSN: 0096-3941."

"8,000 Tons-an-Hour Project in Liberia." 1958. Mining Journal. January 3, 1958. Volume 250, number 6385, page 14.

Elkan, Edward F. 1970. "The Making of a medium to large mine." Bulletin - Geological, Mining and Metallurgical Society of Liberia. 4; Pages 71-75. 1970. Notes: Vol. 4. Geological, Mining and Metallurgical Society. Monrovia, Liberia. Descriptors: Africa; economic geology; evaluation; Liberia; mining geology; West Africa; Economic geology, general. ISSN: 0367-4819.

Elder, T. G. 2002. "Mineral legislation of Liberia." Transactions of the Institution of Mining and Metallurgy, Section B: Applied Earth Science 111 (2002). Abstract: Liberia is underlain predominantly by Archaean and Proterozoic terrain, which is highly prospective, especially for gold and diamonds. Although the country is now governed as a multi-party democracy, Liberia missed out on the boom in mineral exploration that took place elsewhere in sub-Saharan Africa during the 1990s because of civil wars between 1989 and 1995. A new Minerals and Mining Law came into effect in September, 2000, and the provisions of this law for the licensing of exploration and mining are outlined together with the experience of one company, Mano River Resources, which has already negotiated mineral development agreements under the new law. ISSN: 0371-7453.

Ellis, Stephen. "How to Rebuild Africa." Foreign Affairs. Volume 84, no. 5, September-October 2005: page 135. Abstract: This past March, a UN panel revealed that Liberian officials had signed a secret contract with an obscure European company, giving it a virtual monopoly on mining diamonds in the troubled country- even though Liberia has been banned by the UN from selling its diamonds since 2001. The arrangement, it was disclosed, had involved members of the new transitional government operating under the (supposed) scrutiny of a large UN mission. The discovery should not have come as a surprise. Liberia's new government, supposedly a model of national reconciliation, is largely made up of former militia members. During 15 years of war, armed gangs ravaged Liberia, turning it into a classic example of a failed state. Since the fighting stopped in August 2003, the erstwhile warlords have been quick to set aside their differences -- at least when doing so helps them acquire more loot. The mining deal was just one in a long series of similar scandals perpetrated by senior members of the transitional government, who are rapidly signing away their country's future in return for personal financial gain.

Engstrand, Lars G. 1965. "Stockholm natural radiocarbon measurements VI." Radiocarbon. 7; Pages 257-290. 1965. American Journal of Science. New Haven, CT, United States. Abstract: This is a continuation of a series of reports of carbon-14 ages obtained for geologic and archaeologic materials from localities in Sweden and elsewhere. (For reference to the previous report, see this Bibliography Vol. 29, p. 409, Oestlund, H. G.). Descriptors: absolute-age; Africa; Buchanan; C-14; carbon; dates; Europe; Greece; Hoernjafjoerder; Iceland; isotopes; Italy; Liberia; radioactive-isotopes; regional; Scandinavia; Southern Europe; Sweden; Sweden, Italy, Greece, Iceland, Liberia; West Africa; Western Europe. ISSN: 0033-8222.

Ercelawn, Aliya. 2002. "Sea turtles in Liberia : a baseline survey." Tropical Resources. Vol. 21. 2002. p. 17-24. Maps. Descriptors: Sea turtles- Liberia; Reptiles- Liberia; Liberia.

Erickson, Roland I. 1954. Geology of Bomi Hills, Liberia, Africa. Master's thesis, University of North Dakota. Grand Forks, ND, United States. Pages: 68. 1954. Description: iii, 68 leaves: map, plates, charts; 28 cm. Descriptors: Africa; areal geology; Bomi Hill; Liberia; West Africa; Geology-Africa; Geology-Liberia-Tubmanburg. Notes: Typescript. Includes bibliographical references (leaf 68). OCLC: 17255133.

Evans, Jon Michael. 2001. "Gold exploration in tropically weathered terrains: The formation, evolution and geochemistry of lateritic profiles in Liberia and Guinea, West Africa." [Ph.D. dissertation].England: University of Southampton (United Kingdom); 2001. Subjects: Gold, Lateritic, Liberia, Guinea. Geology, Geochemistry. Abstract (Document Summary): Gold-mineralisation at Largor, Liberia, is hosted within a discrete, amphibolite and granitoid bound, E-W trending zone of variably sheared and recrystallised ultramafic-mafic (amphibole+chlorite+/-serpentine+talc) schists. The host exhibits E-W-trending dextral ductile shear deformation and lower amphibolite facies (c.550°C) metamorphism. There is little evidence for pervasive hydrothermal alteration. Two distinct associations of disseminated sulphides and arsenides are developed; pyrrhotite-chalcopyrite is ubiquitous, possibly representing a remobilized primary igneous assemblage; arsenopyrite- niccolite-pyrrhotite is more locally developed, related to a later-stage As- rich fluid. Gold mineralisation is disseminated, occurring dominantly as (syngenetic?) inclusions within amphiboles, with lesser occurrences of host- rock and Au intimately associated with pyrrhotite. Intense tropical weathering has created a residual ferralitic stone-line-type profile comprising of a well developed saprolite, capped by a thin nodular/ gravelly soil and occasionally clayey-silty soils. Detailed textural studies show that weathering commences by the breakdown of primary sulphides and arsenides. Saprolitisation is isovolumetric. Upper saprolite comprises of an open box work of peripheral pseudomorphs, comprising of Fe oxyhydroxide- and kaolinite-rich rinds, developed by incongruent dissolution of amphiboles in the lowermost saprolite and chlorite in the upper saprolite. Partial gibbsitic pseudomorphs form directly during chlorite alteration. Partial kaolinite pseudomorphs after amphibole are occasionally developed in the chlorite-dominated domains of the lower saprolite. Gravelly soils (pisoliths, saprolitic relicts and quartz fragments in a kaolinitic matrix) have formed by the physical collapse of the upper saprolite. The presence of lateritic pisoliths incorporated within soils is strong evidence of a previously dismantled duricrust. More recent truncation down to saprolite levels in some locations has occurred, and residual pisoliths have been incorporated into the soils during saprolite collapse. Subsequent surficial leaching is promoting the removal of Fe, forming residual kaolinite- and quartz-rich clayey-silty soils. Localized interaction with the water-table is promoting Fe remobilization, transforming the upper saprolite into a mottled clay, eventually forming an outcropping secondary pseudo-duricrust. DAI-C 64/01, p. 109, Spring 2003.

"Exploration; country updates." 2001. African Mining. 6; 1, Pages 8, 11, 13, 15, 17, 19. 2001. Descriptors: Africa; Central Africa; Congo; copper ores; diamonds; drilling; East Africa; economics; Ghana; gold ores; Guinea; igneous rocks; kimberlite; Liberia; Mauritania; metal ores; mineral-assemblages; mineral exploration; mineralization; mining; mining-geology; Mozambique; Namibia; plutonic rocks; production; South Africa; Southern Africa; Tanzania; ultramafics; West Africa; Zambia; Zimbabwe; zinc ores.

Fairbain, W.C. 1981. "Diamonds in Liberia." Mining Magazine (London). 144; 1, Pages 40-43. 1981. Mining Journal Ltd.. London, United Kingdom. Abstract: Panorama of the diamond industry in Liberia with a descriptive inventory of deposits, a review of legislation concerning the industry and a table of exports from 1965 to 1979. 1979 exports: 301,000 carats valued at 30 million US dollars. Descriptors: 1965-1979; Africa; diamonds; economic geology; economics; export; export-value; inventory; Liberia; mining legislation; West Africa. ISSN: 0308-6631.

Fairhead, James, et. al., eds., 2003, African-American Exploration in West Africa: Four Nineteenth-century Diaries. Bloomington: Indiana University Press, 2003. 488 pages. Maps. Descriptors: Exploration and discovery; African-Americans- History; Liberia; Guinea.

Fairhead, James and Leach, Melissa. 1998. "Reconsidering the extent of deforestation in twentieth century West Africa." Unasylva (FAO)., v. 49(= no. 192) p. 38-46. Abstract: This article suggests that the extent of deforestation that has occurred in West Africa during the twentieth century is currently being exaggerated. It presents key findings of detailed research into vegetation change over the past century in Côte d'Ivoire, Sierra Leone, Liberia, Ghana, Togo and Benin. ISSN: 0041-6436. FAO: http://www.fao.org/WAICENT/FAOINFO/FORESTRY/UNASYLVA/192/E/192-04E.PDF

Famolu, J. K. 1981. "The Makona family-eutric fluvisols (in Liberia)." Fourth Meeting of the West African Sub-Committee for Soil Correlation and Land Evaluation, Banjul, Gambia, 20-27 October 1979. World Soil Resources Reports (FAO), no. 53. West African Sub-Committee for Soil Correlation and Land Evaluation. Meeting. 4. Banjul (Gambia), 20 October 1979. FAO: Rome (Italy). 1981. Pages 59-61. Notes: 4 references. Descriptors: Alluvial Soils; Liberia. Microfiche No: 8216171-185-E. Related monograph: 216171. Acc. No: 216177; MFN; 216177. FAO Library; ISBN:92-5-101117-6.

Fanfant, R. 1970. "Liberia: Soil Survey: Report to the Government." FAO: Rome. 26 pages, 3 tables, 1 appendix & 1 large map attached. 1970. Descriptors: soil surveys; soil chemicophysical properties; soil fertility; soil types; soil profiles; Liberia. Report No: AGL-UNDP/TA 2845. FAO Library.

Farah, Douglas. 2000. "Liberia Reportedly Arming Guerrillas; Rebel Control of Sierra Leone Diamond-Mining Areas Crucial to Monrovia, Sources Say." Washington Post (June 18, 2000):A.21 ISSN: 0190-8286. Abstract: Taylor launched his revolt against Doe in 1989, then helped Sankoh found the RUF in 1991. Compaore, Taylor and Sankoh, as well as many of their senior commanders, trained at Libya's World Revolutionary Headquarters in the 1980s. The Reagan administration regarded Libya as a primary sponsor of international terrorism and saw Doe as a reliable ally. It poured $500 million in aid into Doe's Liberia and pressured Nigeria and other pro-Western governments to intervene militarily, using Sierra Leone as a base, to fight Taylor. But

Taylor ultimately fought to a draw, signed a cease-fire and won a presidential election in 1997. "The deal was that the RUF would help Taylor 'liberate' Liberia and afterward would provide a base for the RUF to enter Sierra Leone," said [Ibrahim Abdullah]. "When the RUF entered Sierra Leone there was a Burkinabe [Burkina Faso] force under their command that Taylor arranged to send in. All the arms for Taylor and the RUF came from Burkina Faso, and were bought in Ukraine. The payment for all this was diamonds that went through Liberia, Burkina Faso and the Ivory Coast." That basic route still works, intelligence officials said. For months, Western military and intelligence officials have reported Taylor's tacit support for the rebels and friendship with their leaders. But in recent days intelligence officials, diplomats and sources with direct knowledge of RUF activities say his support has become more active and the threat of a wider regional war is growing. These sources say Taylor's recent reinforcement of the rebels is due to his determination to either maintain RUF control over the bulk of Sierra Leone's diamond fields, or back a new RUF escalation of the war. While Taylor acknowledges a friendship and historical ties with RUF leaders, he denies that he is arming the rebels now.

Fayia, A. Kpandel. 2005? "Republic of Liberia: Mining and Investment Opportunities. Ministry of Lands, Mines and Energy. Descriptors: barite; bauxite; diamonds; gold; heavy minerals; hydrocarbons; iron; kyanite; manganese; phosphate. See: http://www.liberiapastandpresent.org/Diversen/INDABA.ppt

Findlay, D. 1998."Boudinage; a key to an organizing principle for the formation of ore deposits." Economic Geology and the Bulletin of the Society of Economic Geologists. 93; 5, Pages 671-682. 1998. Economic Geology Publishing Company. Lancaster, PA, United States. 1998. Descriptors: Africa; Australasia; Australia; boudinage; Broken Hill Deposit; cleavage; Commonwealth of Independent States; diamonds; Dnepropetrovsk, Ukraine; Eastern Goldfields; epigene processes; Europe; folds; foliation; fractures; gems; gold ores; Hamersley Basin; igneous rocks; intrusions; iron ores; Kalgoorlie Deposit; Kimberley, Australia; kimberlite; Krivoy Rog, Ukraine; lead-zinc deposits; Liberia; Lunnon Shoot Deposit; metal ores; mineral deposits,-genesis; mineralization; natural-gas; New South Wales Australia; nickel ores; Nimba Range; petroleum; pipes; plutonic rocks; silver ores; stratiform deposits; structural analysis; Ukraine; ultramafics; veins; West Africa; Western Australia; Western Europe; Economic geology- general deposits; structural geology. References: 82; illustrations, including block diags., sections, sketch maps. ISSN: 0361-0128.

Findlay, D. 1994. "Diagenetic boudinage, an analogue model for the control on hematite enrichment iron ores of the Hamersley Iron Province of Western Australia, and a comparison with Krivoi Rog of Ukraine, and Nimba Range, Liberia." Ore Geology Reviews. 9; 4, Pages 311-324. 1994. Elsevier. Amsterdam, International. 1994. Abstract: A prima facie comparison is made between diagenetic, "sedimentary" boudinage structures at outcrop scales (scales of centimetres to tens of centimetres), and zones of localised stratigraphic thinning (on scales of tens of metres) in beds of the Marra Mamba and Brockman Iron Formations of the Hamersley Iron Province of Western Australia. If the comparison is valid, it suggests that some of the hematite enrichment ores of the province may be diagenetic ores located in necks of diagenetic boudinage structures related to extensional disturbance of the basin when the sequence was only partly consolidated. This interpretation is seen as similar to the consensual supergene metasomatic replacement hypothesis for the origin of the ores in respect of mineral solution-precipitation mechanisms, but differs in respect of important aspects of bulk process, and in their implications for iron ore exploration. A prima facie comparison is also made with the structure locating some ores of the Krivoi Rog region of Ukraine for which a boudinage control has been explicitly described, and with the structure controlling the Nimba Range deposit, Liberia. If such a comparison is valid, then boudinage could account simultaneously for the Proterozoic age of the deposits, the localised stratigraphic thinning, the influx of iron, and the "removal" of silica. Further, on the basis of self-similarity of boudinage structure across scale, region and tectonic regime, and in conjunction with the recognition by others on different grounds that the examples described in the paper may be extrapolated world-wide, boudinage may provide a partial framework within which existing models for the formation of enriched hematite ores of Proterozoic banded iron formations can be adapted. The paper is conceptual and provides no new data. Descriptors: Africa; Australasia; Australia; Brockman Formation; chemically precipitated rocks; Commonwealth of Independent States; controls; diagenesis; Europe; Hamersley Range; hematite; iron formations; iron ores; Krivoy Rog Basin; Liberia; Marra-Mamba-Formation; metal ores; mineral deposits, genesis; Nimba Mountains; oxides; sedimentary processes; sedimentary rocks; sedimentary structures; Ukraine; West Africa; Western Australia. ISSN: 0169-1368.

Firestone Plantations Company. 1959. Firestone Plantations Harbel group, Liberia. [Monrovia, Liberia?]: Firestone Plantations. 1 map; col.; 50 x 55 cm. Descriptors: Rubber- Liberia– Farmington River Valley- Maps; Oil palm- Liberia- Farmington River Valley- Maps. Map Scale 1:75,000. Note(s): Shows Firestone plantation divisions in the Farmington River Valley. Also shows oil palms. Pictorial map. In lower right margin: Harbel, Nickie Kohout, 11-3-59. OCLC: 23382057.

Firestone Plantations Company. 1926. "Map of Liberia Showing Areas Under Development and Districts Already Explored." October 20, 1926. Notes: "Traced from Base Map, Republic of Liberia- 1922." "Traced from original map entitled as above Sept 1941. Scale 1:50,000. Library of Congress, Geography & Map Division.

Firestone Plantations (Misc.) 1926, 1941. Agriculture map file, Du, Farmington and Harbel, aerial view. Various maps, various scales, various dates. "Abrams Aerial Su. Corp." "Tracing from Pan Air, 1941." Descriptors: Africa- Liberia- Agriculture. Library of Congress, Geography & Map Division.

Firestone Rubber Plantations. 1938. "Agriculture." Scale 1:1,800,000. Coverdale & Colpitts. Library of Congress, Geography & Map Division.

Firestone Tire & Rubber Co. 1937. "Liberia: Rubber." Cavalla Group (Region). Scale 1:2,000. Notes: Blueprint. Dummy made for "Liberia Cavalla Group." Library of Congress, Geography and Map Division.

Fitzhugh, E. F, Jr. 1953. "Iron ore at Bomi Hill, Liberia." Economic Geology and the Bulletin of the Society of Economic Geologists. 48; 6, Pages 431-436. 1953. Economic Geology Publishing Company. Lancaster, PA, United States. Abstract: The iron deposit exposed on the crest of Bomi Hill north of Monrovia, Liberia, consists of massive magnetite and hematite. The ore zone overlies granitic gneiss and is overlain locally by amphibolite gneiss and elsewhere by banded iron rich siliceous rocks of probable sedimentary origin. Surface induration and recrystallization of the ore is attributed to the action of meteoric waters. Descriptors: Africa; Bomi Hill; iron; Liberia; metals; mineral deposits, genesis; West Africa. ISSN: 0361-0128.

Fitzhugh, E. F, Jr. 1951. "Iron ore at Bomi Hill, Liberia." Economic Geology and the Bulletin of the Society of Economic Geologists. 46; 7, Pages 800. 1951. Economic Geology Publishing Company. Lancaster, PA, United States. Descriptors: Africa; Bomi hill; iron; Liberia; metals; West Africa. Geol. Soc. Am., B. v. 62, No. 12, pt. 2, p. 1437. ISSN: 0361-0128.

Folger, D. W.; Irwin, B. J.; McCullough, J. R.; Rowland, R. W. and Polloni, C. F. 1991. Map showing free-air gravity anomalies off the southern coast of west-central Africa; Liberia to Ghana. Serial Miscellaneous Field Studies Map. Report number: MF-2098-E. Date: January 1, 1990. (1 sheet). U. S. Geological Survey, Reston, VA, United States. Map Scale: 1:500,000; Map Type: gravity survey map. Annotation: Lat 2 degrees N to 10 degrees N, long 2 degrees E to 11 degrees W. Scale 1:500,000 (1 inch = about 8 miles). Sheet 50 by 28 1/2 inches. Descriptors: Africa; free-air anomalies; geophysical methods; geophysical surveys; Ghana; gravity anomalies; gravity methods; gravity survey maps; Guinea; Ivory Coast; Liberia; maps; surveys; USGS; West Africa.

Folger, David W. 1990. Map showing free-air gravity anomalies off the southern coast of west-central Africa: Liberia to Ghana. Map showing free air gravity anomalies off the southern coast of west-central Africa. Free air gravity anomalies off the southern coast of west-central Africa. United States. Defense Mapping Agency; Geological-Survey-(U.S.) Reston, VA; Denver, CO, United States. 1 map; 59 x 96 cm., on sheet 72 x 127 cm., folded in envelope 30 x 24 cm. Scale 1: 1,500,000; Lambert azimuthal equal area proj. (W 11-E 2/N 10-N 2). Series: Miscellaneous field studies; map MF-2098-E. Notes: Includes text and location map. Includes bibliographical references. Descriptors: gravity anomalies; Atlantic Coast; Africa Maps. OCLC: 23184311; USGS Library: M(200) MF no.2098-E.

Fomin, Yu M. and Melnikov, F. P. 2000. Svyaz' kimberlitovogo magmatizma s megablokami drevnikh platform. Translated: Relations of kimberlite magmatism with ancient platform megablocks. Vestnik Moskovskogo Universiteta, Seriya 4, Geologiya. 2000; 3, Pages 56-65. 2000. Language: Russian. Descriptors: Africa; Anabar-Shield; Asia; Commonwealth-of-Independent-States; cratons; crust; diamonds; genesis; igneous-rocks; Kaapvaal-Craton; kimberlite; Liberia; Liberian-Shield; magmatism; platforms; plutonic-rocks; Russian-Federation; Siberian-Platform; Southern Africa; ultramafics; West Africa; Yakutia-Russian-Federation. References: 18; illustrations, including sketch map. ISSN: 0579-9406

Force, Eric R. 1983. "Geology of Nimba County, Liberia." USGS Bulletin. Report number: B 1540. January 1, 1983. Pages (monograph): 27. U. S. Geological Survey, Reston, VA, United States. 27 pages: illustrations, maps; 24 cm. Abstract: Precambrian rocks of Nimba County, grouped in two initially separated tectonic terranes, the Nimba Block and the Gbedin-Kahnple Block, were juxtaposed, redeformed, metamorphosed, and then over thrust by rocks of a third, younger Precambrian terrane. Notes: Bibliography: p. 25-27. Descriptors: Africa; areal geology; block structures; explanatory text; faults; folds; Gbanka Quadrangle; geologic maps; isoclinal folds; Liberia; maps; metamorphic rocks; mineral resources; Nimba County; Nimba Mountains; over thrust faults; Precambrian; Sanokole Quadrangle; tectonics; USGS; West Africa; Zwedru

Quadrangle. Number of references: 31. Illustrations, including 3 tables, geological sketch map. ISSN: 8755-531X; LCCN: 82-600081; OCLC: 8283004.

Force, Eric R. 1981. "Geology of Nimba County, Liberia." Liberian Geological Survey. United States; Agency for International Development; Geological Survey (U.S.). [Reston, Va.?]: U.S. Dept. of the Interior, Geological Survey. Description: ii, 37 p., 3 p. of plates: ill., maps; 28 cm. Series: Project report: Liberia investigations. Geological Survey (U.S.); LI-86; Variation: Project report (Geological Survey (U.S.)); LI-86. Descriptors: Geology- Liberia- Nimba County. Notes: Cover title. "Prepared in cooperation with the Liberian Geological Survey, under the auspices of the Agency for International Development." Bibliography: p. 34-37. OCLC: 12261759.

Force, Eric R. and Beikman, H. M. 1977. "Geologic map of the Zwedru Quadrangle, Liberia." Liberian Geological Survey. Miscellaneous Investigations Series Map. Report number: I-0777-D. US Geological Survey, Reston, VA, United States. January 1, 1977. Map Type: geologic map; color map; 44 x 55 cm. on sheet 73 x 105 cm. folded in envelope 30 x 24 cm. Scale 1:250,000 (1 inch = about 4 miles). Sheet 29 by 41 inches. Descriptors: Africa; areal geology; geologic maps; Liberia; maps; USGS; West Africa; Zwedru Quadrangle. Notes: "Hotines rectified skew orthomorphic projection and rectified coordinates." "Prepared by the U.S. Geological Survey and the Liberian Geological Survey under the joint sponsorship of the Government of Liberia and the Agency for International Development, U.S. Department of State". Includes bibliography. ISSN: 0160-0753. OCLC: 3593200

Force, Eric R. and Beikman, Helen M. 1974. "Geologic map of the Zwedru Quadrangle, Liberia." Open-File Report. January 1, 1974. Report number: OF 74-0307. Pages: 12 (1 sheet). U. S. Geological Survey, Reston, VA, United States. 12 leaves: folded map; 27 cm. Annotation: 1 sheet, scale 1:250,000 (1 inch = about 4 miles). Descriptors: Africa; areal geology; geologic maps; Liberia; maps; USGS; West Africa; Zwedru Quadrangle. Notes: Map in pocket. Prepared under the auspices of the Govt. of Liberia and the Agency for International Development, U.S. Dept. of State. Transmittal sheet dated November 11, 1974. Bibliography: leaves 11-12. ISSN: 0196-1497.

Force, Eric R. and Berge, W. J. 1977. "Geologic map of the Gbanka Quadrangle, Liberia." Liberian Geological Survey. Miscellaneous Investigations Series Map. Report number: I-0776-D. U. S. Geological Survey, Reston, VA, United States. January 1, 1977. Monograph color map; 44 x 44 cm. on sheet 74 x 106 cm. folded in envelope 30 x 24 cm. Map Scale 1:250,000 (1 inch = about 4 miles). Sheet 29 by 42 inches. Descriptors: Africa; areal geology; Gbanka Quadrangle; geologic maps; Liberia; maps; USGS; West Africa. Notes: "Hotines rectified skew orthomorphic projection and rectified coordinates." "Prepared by the U.S. Geological Survey and the Liberian Geological Survey under the joint sponsorship of the Government of Liberia and the Agency for International Development, U.S. Department of State. Includes bibliography. ISSN: 0160-0753; OCLC: 3594342.

Force, Eric R. and Berge, J. W. 1977. "Geologic map of the Sanokole Quadrangle, Liberia." Miscellaneous Investigations Series Map. Report number: I-0774-D. Date: January 1, 1977. U. S. Geological Survey, Reston, VA, United States. Map Type: geologic map; color map; 44 x 44 cm. on sheet 74 x 105 cm. folded in envelope 30 x 24 cm. Scale 1:250,000 (1 inch = about 4 miles). Sheet 29 by 42 inches. Descriptors: Africa; areal geology; geologic maps; Liberia; maps; Sanokole

Quadrangle; USGS; West Africa. Notes: "Hotines rectified skew orthomorphic projection and rectified coordinates." "Prepared by the U.S. Geological Survey and the Liberian Geological Survey under the joint sponsorship of the Government of Liberia and the Agency for International Development, U.S. Department of State." Includes bibliography. ISSN: 0160-0753.

Force, Eric R. and Berge, J. W. 1974. "Geologic map of the Sanokole Quadrangle, Liberia. Geology of the Sanokole quadrangle." Open-File Report. Report number: OF 74-0304. Pages (monograph): 17 pages (1 sheet). January 1, 1974. U. S. Geological Survey, Reston, VA, United States. 17 leaves: folded map; 27 cm. Map scale 1:250,000 (1 inch = about 4 miles). Descriptors: Africa; areal geology; geologic maps; Liberia; maps; Sanokole Quadrangle; USGS; West Africa. Notes: Map in pocket. Prepared under the auspices of the Govt. of Liberia and the Agency for International Development, U.S. Dept. of State. Bibliography: leaves 16-17. ISSN: 0196-1497.

Force, Eric R. and Dunbar, J. D. N. 1974. "Geologic map of the Gbanka Quadrangle, Liberia; Geology of the Gbanka quadrangle, Liberia." Open-File Report. Report number: OF 74-0306. Pages (monograph): 14 (1 sheet). Date: January 1, 1974. U. S. Geological Survey, Reston, VA, United States. 14 leaves: folded maps; 27 cm. Annotation: 1 sheet, scale 1:250,000 (1 inch = about 4 miles). Descriptors: Africa; areal geology; Gbanka Quadrangle; geologic maps; Liberia; maps; USGS; West Africa. Geologic maps. Notes: Map in pocket. Prepared under the auspices of the Govt. of Liberia and the Agency for International Development, U.S. Dept. of State. Transmittal sheet dated Dec. 13, 1974. Bibliography: leaf 14. ISSN: 0196-1497.

Francis, C. A.; Massaquoi, W. K.; Beebe, J.; Davidson, D. J.; Massaquoi, R. C. and Mulbah, C. K. 1995. "Designing an integrated cropping systems research program: Central Agricultural Research Institute (CARI), Liberia." Journal of Sustainable Agriculture 5, no. 3 (1995) pages 147-168. Abstract: The Central Agricultural Research Institute (CARI) is located in the humid forest zone near Suakoko, Bong County, Liberia, with national responsibility for basic and applied agricultural research. A new initiative will integrate technical research disciplines with extension workers and farmers into an effective cropping systems research (CSR) and outreach effort. The objective is to shift the focus away from disciplines and the experiment stations and to incorporate appropriate component technologies into sustainable, environmentally sound, and economically viable prototype systems that can be tested on farms. The process involves setting clear short- and long-term goals in the institute, and elaborating a program design within the Liberian farmer's resource context. Evaluation of the program's success depends on measuring adoption of practices and systems, increases in farm productivity and economic return, and ecological sustainability. The program can serve as a model for other countries in the humid tropics with research resource constraints and organizational challenges. Descriptors: Arable land; Agricultural research and extension; Liberia.

Fumbah, D. F. and Loetsch, D. 1974. "National forests inventory of Liberia." IN: FAO/SIDA Training Course on Forest Inventory, Ibadan (Nigeria), 12 August 1974. FAO: Rome (Italy). Forest Resources Div. Report on the Second FAO/SIDA Training Course on Forest Inventory, Ibadan, Nigeria, 12 August - 13 September 1974, pages 188-198. Descriptors: forest inventories; classification; species; surveying; maps. M Report No: FOR--TF-RAF-73(SWE). FAO Library.

Furbay, Elizabeth Jane Dearmin. 1943. Top hats and tom-toms, by Elizabeth Dearmin Furbay. Chicago, New York, Ziff-Davis Pub. Co. [1943]. Description: Book 6 p. l., 307 p. ill., plates. 21 cm. Descriptors: Liberia Description and travel. University of Wisconsin- American Geographical Society Collection. Call Number: DT626 .F8 1943.

Furon, Raymond. 1969. "Introduction à la Géochronologie de l'Afrique." Lexicon de Stratigraphie: Afrique. Volume IV, Fasc. 13, 1-73. Union Internationale des Sciences Géologiques, Commissions de Stratigraphie, Sous-Commission du Lexique Stratagraphique, Centre National de la Recherche Scientifique (CNRS), Paris. "4. Le bouclier Ivoirien (ou Éburnéen): Liberia; Guinée; Sierra Leone; Côte d'Ivoire; Mali; Niger; Haute Volta; Ghana."

Gårdlund, Torsten. 1968. Lamco in Liberia. Stockholm, Grängesberg Co. 134 pages, illustrations. Descriptors: Technical assistance, Swedish- Liberia; Liberia- Economic conditions. Notes: Cover title. Reproduced from typewritten copy. Translation of "Lamco i Liberia" (see below). Bibliography: pages 129-134. OCLC: 15726708.

Gårdlund, Torsten. 1967. Lamco i Liberia. Stockholm, Almqvist & Wiksell. Language: Swedish. 149 pages, illus. (6) l. of plates. 21 cm. Descriptors: Technical assistance, Swedish- Liberia; Liberia- Economic conditions. Notes: Bibliography: pages 146-149. LCCN: 68-105175; OCLC: 20498078.

Garnett, T. and Utas, C. 2002. "The Upper Guinea Heritage. The status of nature conservation in Liberia and Sierra Leone." International Union for Conservation of Nature and Natural Resources, Rue Mauverney 28 Gland CH-1196 Switzerland. 60 pages. 2002. Descriptors: tropical environment; rain forests; books; population growth; resource utilization; logging; mining; social environment; conservation; Liberia; Sierra Leone; Cote d'Ivoire; Ghana; Guinea. Abstract: The two neighbouring countries, Liberia and Sierra Leone, are situated in the heart of the Upper Guinea forest region. This region is one of the earths most biologically diverse and was originally covered by a continuous block of dense tropical rainforest, ranging from Guinea south through Sierra Leone and Liberia to Ivory Coast and Ghana. Much of this forest cover has already been lost and what little remains is under serious threat from commercial activities, such as logging, mineral mining and the subsistence activities of an ever growing population. Civil conflicts are also taking their toll. The status of nature in the two countries will be depicted using case studies and pictures. ISBN: 9075909063. See: http://www.iucn.org

Gaskill, G. 1948. "Our Stake in Liberia." American Magazine (Springfield, Ohio). Volume 146, July 1948, pages 58-60. The same abridged, with the title: "Liberia: A New Frontier," in Reader's Digest, volume 53, October 1948, pages 94-98. Abstract: Description of Liberia with special reference to the new Stettinius enterprise, the Liberia Company, and Firestone.

Gatter, Wulf. 1986. "The Coastal Wetlands of Liberia. ICBP Study Report No.26. 44 pages, 10 figs.

Gazetteer of Liberia: names approved by the United States Board on Geographic Names. Microfiche. 3rd ed., July 1997. 1997. Washington, DC: United States Board on Geographic Names. USGPO Superintendent of Documents number: D 5.319:L 61/2.

Geiger, Luther and Nettleton, W. D. 1979. "Properties and geomorphic relationships of some soils of Liberia." In: Proceedings of the 44th annual meeting, Soil Science Society of America. Ellis, Roscoe Jr. (editor). Soil Science Society of America Journal. 43; 6, Pages 1192-1198. 1979. Soil Science Society of America. Madison, WI, United States. Conference: 44th Annual Meeting. Detroit, MI, United States. November 30-December 5, 1980. Descriptors: Africa; clay-soils; environment; geochemistry; geomorphology; ion-exchange; Liberia; Oxisols; properties; soil surveys; soils; surveys; tropical-environment; Ultisols; vegetation; West Africa; soil science, soil classification; soil properties; geomorphology; soil chemistry; soil mineralogy; weathering (geology); Saprolites. Abstract: Plinthic Paleudults of the clayey-skeletal family are on the highest uplands. Compared to the other Paleudults, they are more clayey, and have higher amounts of extractable iron, and contain high amounts of ironstone concretions. The argillic horizons have 0.3 meq of Ca 2 + and Mg 2 + or less per 100 g of soil. Because of the gravel, these soils are best suited to rubber or cashews. The loamy family of Plinthic Paleudults and the clayey family of Typic Paleudults occupy lower erosional uplands. They are similar to the higher, clayey-skeletal Paleudults but are slightly higher in bases in the lower horizons. Both are well suited to production of oil palm, coffee, rubber, and cashews. The Typic Tropopsamments and Aquoxic Dystropepts of the clayey family occupy low stream terraces. Both soils have more weatherable minerals and a more favorable base status than the Paleudults. The Dystropepts are best suited to oil palm production. The Tropopsamments are well suited to cacao coffee, oil palm, and rubber. The Paleudults studied have the chemical properties of the Oxisols. They have a much lower CEC per unit of clay, and lower base saturation than apparently is intended for Ultisols. We propose that a new subgroup, Aquoxic Paleudults, be provided for these soils. References: 13; 4 tables, block diag., sketch map. ISSN: 0361-5995.

Genevray, J. 1952. "Élements d'une Monographie d'une Division Administrative Libérienne (Grand Bassa county)." Dakar: IFAN. 135 pages. Mémoires de l'Institut Français d'Afrique Noire. Number 21. Abstract: Treatise on a region of 9,000 square kilometers along the central coast and reaching back into the interior of Liberia. Subjects covered include physical setting and climate, origins and history of the landings and settlements of colonists from America, which are still limited to the towns of the coastal strip, anthropometry and habitat of the indigenous tribes, Bassa, Kru and Vai, the Liberian administrative system, religious practices of Liberians and of the partially Christianized- Protestant Africans, and the language of the Bassa. An appendix is an analysis of observations of over a thousand cases of yaws among bare footed natives- a disease, says the author, which never attacks those who wear shoes. OCLC: 823237.

"Geographic map of the Bopolu Quadrangle, Liberia." 1973. Miscellaneous Investigations Series Map. Report number: I-0772-A. January 1, 1973. U. S. Geological Survey, Reston, VA, United States; Liberian Geological Survey, Liberia. Annotation: Lat 7 degrees to 8 degrees, long 10 degrees to 11 degrees. Scale 1:250,000 (1 inch = about 4 miles). Sheet 28 by 29 inches. Descriptors: Africa; Bopolu Quadrangle; geographic; Liberia; maps; USGS; West Africa; Geologic maps. ISSN: 0160-0753.

"Geographic map of the Buchanan Quadrangle, Liberia." 1973. Miscellaneous Investigations Series Map. Report number: I-0778-A. January 1, 1973. U. S. Geological Survey, Reston, VA, United States; Liberian Geological Survey, Liberia. Annotation: Lat 5 degrees to 6 degrees, long 9 degrees to 10 degrees. Scale 1:250,000 (1 inch = about 4 miles). Sheet 26 by 29 inches. Descriptors: Africa; Buchanan Quadrangle; geographic; Geologic maps; Liberia; maps; USGS; West Africa. ISSN: 0160-0753.

"Geographic map of the Gbanka Quadrangle, Liberia." 1972. Miscellaneous Geologic Investigations Map. Miscellaneous investigations series (Geological Survey (U.S.)); map I-776-A. U. S. Geological Survey, Reston, VA, United States. January 1, 1972. Map Scale: 1:250,000. Annotation: Lat 6 degrees to 7 degrees, long 9 degrees to 10 degrees. Scale 1:250,000 (1 inch = about 4 miles). Sheet 21 by 29 inches. Descriptors: Africa; areal geology; Gbanka Quadrangle; geographic; Liberia; maps; USGS; West Africa; Geologic maps. ISSN: 0375-8001.

"Geographic map of the Harper Quadrangle, Liberia." 1973. Miscellaneous Investigations Series Map. January 1, 1973. Report number: I-0780-A. U. S. Geological Survey, Reston, VA, United States; Liberian Geological Survey, Liberia. Annotation: Lat 4 degrees to 5 degrees, long 7 degrees to 9 degrees. Scale 1:250,000 (1 inch = about 4 miles). Sheet 28 1/2 by 31 1/2 inches. Descriptors: Africa; geographic; Geologic maps; Harper Quadrangle; Liberia; maps; USGS; West Africa. ISSN: 0160-0753.

"Geographic map of the Juazohn Quadrangle, Liberia." 1973. Miscellaneous Investigations Series Map. January 1, 1973. Report number: I-0779-A. U. S. Geological Survey, Reston, VA, United States; Liberian Geological Survey, Liberia. Map Scale: 1:250,000. Annotation: Lat 5 degrees to 6 degrees, long 7 degrees20#PR to 9 degrees. Scale 1:250,000 (1 inch = about 4 miles). Sheet 29 by 33 inches. Descriptors: Africa; geographic; Juazohn Quadrangle; Liberia; maps; USGS; West Africa. ISSN: 0160-0753.

"Geographic map of the Monrovia Quadrangle, Liberia." 1973. Miscellaneous Investigations Series Map. Report number: I-0775-A. January 1, 1973. U. S. Geological Survey, Reston, VA, United States; Liberian Geological Survey, Liberia. Annotation: Lat 6 degrees to 7 degrees, long 10 degrees to 11 degrees. Scale 1:250,000 (1 inch = about 4 miles). Sheet 28 1/2 by 30 inches. Descriptors: Africa; geographic; Liberia; maps; Monrovia Quadrangle; USGS; West Africa; Geologic maps. ISSN: 0160-0753.

"Geographic map of the Sanokole Quadrangle, Liberia." 1973. Miscellaneous Investigations Series Map. January 1, 1973; other date: January 1, 1974. Report number: I-0774-A. U. S. Geological Survey, Reston, VA, United States; Liberian Geological Survey, Liberia. Annotation: Lat 7 degrees to 8 degrees, long 8 degrees to 9 degrees. Scale 1:250,000 (1 inch = about 4 miles). Sheet 21 by 29 inches. Descriptors: Africa; geographic; Liberia; maps; northeast; Sanokole Quadrangle; USGS; West Africa. ISSN: 0160-0753.

"Geographic map of the Voinjama Quadrangle, Liberia." 1973. Miscellaneous Investigations Series Map. Report number: I-0771-A. January 1, 1973. U. S. Geological Survey, Reston, VA, United States; Liberian Geological Survey, Liberia. Map Scale: 1:250,000. Annotation: Lat 8 degrees to 9 degrees, long 9 degrees to 10 degrees. Scale 1:250,000 (1 inch = about 4 miles). Sheet 27 by 29 inches. Descriptors: Africa; geographic; Geologic maps; Liberia; maps; USGS; Voinjama Quadrangle; West Africa. ISSN: 0160-0753.

"Geographic map of the Zorzor Quadrangle, Liberia." 1972. U. S. Geological Survey, Reston, VA, United States. Miscellaneous Geologic Investigations Map. Other Titles: Zorzor, Liberia, geographic. Report number: I-0773-A. January 1, 1972. Geological Survey, Reston, VA., United States; Liberian Geological Survey, Liberia. Map Scale: 1:250,000. Annotation: Lat 7 degrees to 8 degrees, long 9 degrees to 10 degrees. Scale 1:250,000 (1 inch = about 4 miles). Sheet 21 by 29 inches. Descriptors: Africa; areal geology; geographic; Liberia; maps; USGS; West Africa; Zorzor Quadrangle; Geologic maps. Notes: 1 map: color; 44 x 44 cm. folded in envelope 30 x 24 cm. Relief shown by shadings and spot heights. Envelope Zorzor, Liberia, geographic. In upper margin: Republic of Liberia, Ministry of Lands and Mines. Includes map of administrative boundaries and sheet index map. ISSN: 0375-8001.

"Geographic map of the Zwedru Quadrangle, Liberia." 1973. Miscellaneous Investigations Series Map. Report number: I-0777-A. January 1, 1973. U. S. Geological Survey, Reston, VA, United States; Liberian Geological Survey, Liberia. Annotation: Lat 6 degrees to 7 degrees, long 7 degrees to 9 degrees. Scale 1:250,000 (1 inch = about 4 miles). Sheet 25 1/2 by 29 inches. Descriptors: Africa; geographic; Liberia; maps; USGS; West Africa; Zwedru Quadrangle. ISSN: 0160-0753.

"Geography, Resources and Inhabitants of Liberia." 1905. Smithsonian Institution. Annual Report. Pages 247-264.

Geological exploration and resources appraisal, AID project #669-11-210-071 with Geological Survey, Bureau of Natural Resources and Surveys, Government of Liberia: progress to date report, 1 July 1965 to 31 December 1966. Other Titles: Geological exploration and resources appraisal ... Liberia. 1966. US Geological Survey and the Liberian Geological Survey. 21 leaves in various foliations: maps; 27 cm. Descriptors: Geological-mapping-Liberia. USGS Library; OCLC: 6374271

George, Richard P., Jr. and Sims, Richard H. 1993. "Eastern Venezuela Basin's post-Jurassic evolution as a passive transform margin basin." In: AAPG SVG international congress exhibition; abstracts. AAPG Bulletin. 77; 2, Pages 320. 1993. American Association of Petroleum Geologists. Tulsa, OK, United States. 1993. Conference: AAPG/SVG international congress/exhibition. Caracas, Venezuela. March 14-17, 1993. Descriptors: Africa; Antilles; Atlantic Ocean; Blake Plateau; Caribbean region; Caribbean Sea; Ceara Basin; Cenozoic; compression tectonics; Cretaceous; en-echelon folds; evolution; faults; folds; Guianas; Lesser Antilles; Liberia; lithofacies; Mesozoic; Nigeria; North American Atlantic; North Atlantic; Northwest Atlantic; Paleogene; passive margins; Piaui Basin; reconstruction; Serrania del Interior; South America; strike slip faults; tectonics; Tertiary; thickness; transform faults; Trinidad; Trinidad and Tobago; Venezuelan Basin; West Africa; West Indies; Solid earth geophysics; Structural geology. ISSN: 0149-1423.

Gershoni, Yekutiel. 1987. Black Colonialism: the Americo-Liberian Scramble for the Hinterland. Boulder: Westview Press, 1985. Westview special studies on Africa. Descriptors: Historical geography- Liberia.

Gershoni, Yekutiel. 1987. The Drawing of Liberian Boundaries in the Nineteenth Century: Treaties with African Chiefs Versus Effective Occupation. Boston, 1987. Series: The International Journal of African Historical Studies. Vol. 20, no. 2 pages 293-307. Descriptors: Boundaries - Historical geography – Liberia.

Ghering, W. 1966. West Liberia: Area Between Mano River and Bomi Hill. May 1966. Map. Scale: 1:125,000. Notes: "Roads and railroads roughly sketched in!" USGS Library Map Collection

Gillis, M. 1988. "West Africa: resource management policies and the tropical forest." In: Public policies and the misuse of forest resources. Pages 299-351. Cambridge University Press, for World Resources Institute. Abstract: Examines the role of public policy in deforestation in each of four West African countries: Liberia, Ivory Coast, Ghana, and Gabon. Provides an overview of forest resources, deforestation, and international trade in tropical hardwoods for the entire region, and on a country-by-country basis. Patterns of property rights and foreign investment in each nation are addressed, as well as the national benefits these countries have derived from forest utilization and government capture of timber rents. Focuses upon reforestation and forest concessions policies, respectively, and examine the impact of forest-based industrialization policies. Finally, the impact of non-forestry policies on tropical forest utilization in each of the four nations is considered. Notes: Special Features: 1 graph, 49 references, 22 tables.

Gnielinski, Stefan von. 1972. Liberia in maps. London: University of London Press, 1972. 111 pages: ill., maps; 29 cm. Notes: Bibliography: p. 109-111. Contents: Soils, pages 18-19; Drainage, pages20-21. ISBN: 0340158042; LCCN: 73-158263; OCLC: 628993.

Gnielinski, Stefan. 1966. "Some thoughts about the geology of Liberia." The Liberian Naturalist. September 1966.

Gokhberg, M. B.; Gufeld, I. L.; Rozhnoy, A. A.; Marenko, V. F.; Yampolsky, V. S. and Ponomarev, E. A. 1989. "Study of seismic influence on the ionosphere by super long-wave probing of the Earth-ionosphere waveguide." In: Seismoelectromagnetic effects. Parrot, M. J. S. (editor); Johnston, M. J. S. (editor). Physics of the Earth and Planetary Interiors. 57; 1-2, Pages 64-67. 1989. Elsevier. Amsterdam, Netherlands. Conference: IUGG XIX general assembly symposium on Seismoelectromagnetic effects. Vancouver, BC, Canada. August 9-22, 1989. Descriptors: Africa; anomalies; Asia; disturbances; earthquakes; effects; electrical field; Hindu Kush; Indian Ocean Islands; ionosphere; Liberia; Mascarene Islands; Omsk; precursors; radio wave methods; Reunion; seismology; West Africa; Solid-earth-geophysics; Seismology. References: 4; illustrations, ISSN: 0031-9201.

"Gold in Liberia." 1903. Liberia Bulletin (American Colonization Society, Washington, DC). Volume 22, pages 45-48.

Goldring, D. C. 1991. "Significance of pre- or syntectonic origin for certain iron ores hosted in banded iron formation." Institution of Mining and Metallurgy, Transactions, Section B: Applied Earth Science. 100; Pages B148-B158. 1991. Institution of Mining and Metallurgy. London, United Kingdom. 1991. Descriptors: Africa; Brazil; chemically precipitated rocks; D'Idjil; deformation; economic geology; iron formations; iron ores; Kedia; Liberia; M'Haoudat; Mauritania; metal ores; metamorphism; sedimentary rocks; South America; syntectonic processes; Venezuela; West

Africa; Economic geology; geology of ore deposits. Illustrations; References: 48, including 4 tables, sections, sketch maps. ISSN: 0371-7453.

Goudarzi, G. H. 1967. "Geology and Mineral Resources of Liberia, a Reconnaissance." Washington, D.C. Reprinted 1970, U.S. Government Printing Office, 104 pages. Library of Congress.

Grand Bassa County. 1977. Liberian Cartographic Service. [Monrovia]: The Service. 1 map: photocopy; 66 x 77 cm. Descriptors: Grand Bassa County (Liberia)- Maps. Map Info: Scale 1:250,000. Notes: Relief shown by hachures. Includes source note and location map. "Series no. L.C.S. 24-77." Responsibility: prepared by Liberian Cartographic Service, Ministry of Lands & Mines, August 1977. LCCN: 85-690721; OCLC: 12836352.

Grand Gedeh County. 1975. Liberian Cartographic Service. [Monrovia]: Liberian Cartographic Service. 1 map: photocopy; 80 x 77 cm. Descriptors: Grand Jide County (Liberia)- Maps. Scale 1:250,000; (W 9000'--W 7015'/N 6030'--N 4045'). Notes: Includes source note and location map. "Series no. LCS-26-75." LCCN: 85-690929; OCLC: 12942473.

Grängesbergsbolaget. 1961. [Liberia issue]. Stockholm: Grängesbergsbolaget. Other Titles: Malm. 52 pages: illustrations; 27 cm. Language: English. Descriptors: iron ores- Liberia- Nimba, Mount Region; iron mines and mining- Liberia- Nimba, Mount Region; iron industry and trade- Liberia- Nimba, Mount Region; industries- Liberia. Named Corporation: LAMCO. Notes: Title derived from introductory paragraph on inside front cover. This special Liberia issue published as Årgang 10, Nr 1 of Malm, the quarterly magazine of the Grängesberg Company. Original articles appear in Swedish with English translations of photo captions. An English translation of most of the text is bound in with its own page numbering (12 p.). Traces the development of the Lamco (Liberian American-Swedish Minerals Company) Joint Venture. OCLC: 29228336.

Greenhalgh, Peter. 1985. West African diamonds 1919-1983; an economic history. Manchester Univ. Press. Manchester, United Kingdom. Pages: 306. 1985. Notes: "Expanded and revised version of [the author's] Ph. D. thesis at the Centre of West African Studies, University of Birmingham" Includes index. Bibliography: pages 291-297. Descriptors: Africa; diamonds; economic geology; economics; Ghana; Guinea; history; industry; Ivory Coast; Liberia; Sierra Leone; West Africa. References: 228; illustrations, including 10 tables, sketch maps. ISBN: 0719017637; OCLC: 11623274.

Gresham, G. E. 1990. Secondary Wood Processing in Liberia, Cote d'Ivoire, Ghana and Nigeria. Background paper. Performer: United Nations Industrial Development Organization, Vienna (Austria). 9 February 1990. 54p. Notes: Presented at Global Preparatory Meeting for the Consultation on the Wood and Wood Products Industry (2nd), Nairobi, Kenya, April 24-27, 1990. Descriptors: industries; Liberia; Ivory Coast; Ghana; Nigeria; forestry; Economic-conditions; Benefit-cost-analysis; Marketing; wood-products; Africa; developing-countries. Abstract: The report examines the secondary wood processing industry in Liberia, Ivory Coast, Ghana and Nigeria. It covers forestry, forest product processing, financial aspects, economic conditions, cost-benefit analysis, training, marketing, and regional cooperation. NTIS Number: PB90268665XSP.

Griethuysen, H. V. van. 1970. "Mineral exploration of Wm. H. Muller & County in East Liberia." Bulletin - Geological, Mining and Metallurgical Society of Liberia. 4; Pages 88-95. 1970. Notes: Vol. 4. Geological, Mining and Metallurgical Society. Monrovia, Liberia. Descriptors: Africa; east; economic geology; exploration; history; Liberia; mineral exploration; mineral resources; West Africa; economic geology, general. Illustrations: geological sketch map. ISSN: 0367-4819.

Gromme, Sherman and Dalrymple, G. B. 1972. "K-Ar Ages and Paleomagnetism of Dikes in Liberia." Eos, Transactions, American Geophysical Union. 53; 11, Pages 1130. 1972. American Geophysical Union. Washington, DC. Descriptors: absolute-age; Africa; age; basalts; dates; diabase; dikes; geochronology; igneous-rocks; intrusions; K-Ar; Liberia; northwest; paleomagnetism; plutonic rocks; pole positions; volcanic rocks; West Africa; geochronology. ISSN: 0096-3941.

De Groot, Ir. P. F. 1936. Provisional map of Liberia NW of the St. Paul River. The Hague; [s.n.]. Description: 1 map; 125 x 110 cm.; Scale 1:200,000. Descriptors: Liberia- Maps. Notes: Relief shown by spot heights. Surveyed and drawn for the Holland Syndicate by P. F. de Groot. Survey: Compass-time estimate method. Fieldwork May 1934 till May 1935 & November 1935 till May 1936. OCLC: 38735832.

Gruss, H. 1973. "Itabirite iron ores of the Liberia and Guyana shields." In: Genesis of Precambrian iron and manganese deposits--Genese des formations precambriennes de fer et de manganese. Earth Science (Paris) = Sciences de la Terre (Paris). 9; Pages 335-359. 1973. UNESCO. Paris, France. Language: English; Summary Language: French. Descriptors: Africa; Bomi Hills; Bong Range; Cerro Bolivar; chemically precipitated rocks; economic geology; El Pao; genesis; grade; Guyana-Shield; iron formations; iron ores; itabirite; Liberia; metal ores; metamorphic rocks; metamorphism; mineral deposits, genesis; Nimba; ore deposits; Precambrian; processes; reserves; San-Isidro; sedimentary rocks; sedimentation; South America; supergene processes; Venezuela; West Africa; Economic geology of ore deposits. Illustrations, including geologic. sketch maps. ISSN: 0070-7910.

Gupta, R. S. 1981. "Multi-site streamflow simulation of Saint John River in Liberia." Journal of the Institution of Engineers. India. Part CH. Chemical Engineering Division. 62; 1, Pages 1-6. 1981. Editor, Institution of Engineers (India). Calcutta, India. Descriptors: Africa; data processing; digital simulation; hydrogeology; hydrology; Liberia; mathematical models; river discharge; Saint John River; stochastic processes; surveys; West Africa. ISSN: 0020-3351.

Haggerty, Stephen E. 1983. "A freudenbergite-related mineral in granulites from a kimberlite in Liberia, West Africa." Neues Jahrbuch fuer Mineralogie. Monatshefte. 1983; 8, Pages 375-384. 1983. E. Schweizerbart'sche Verlagsbuchhandlung. Stuttgart, Federal Republic of Germany. Descriptors: Africa; chemical composition; crust; crystal zoning; electron probe data; freudenbergite; granulites; igneous rocks; ilmenite; inclusions; kimberlite; Liberia; lower crust; metamorphic rocks; mineralogy; minerals; nesosilicates; orthosilicates; oxides; paragenesis; perovskite; petrology; plutonic rocks; rutile; silicates; titanite; ultramafics; West Africa; xenoliths; mineralogy of non silicates; igneous and metamorphic petrology. References: 20; illustrations, including 10 anal., 1 table. ISSN: 0028-3649.

Haggerty, Stephen E. 1983. "Oxide-silicate reactions in lower crustal granulites from Liberia, West Africa." In: The Geological Society of America, 96th annual meeting. Abstracts with Programs - Geological Society of America. 15; 6, Pages 589. 1983. Geological Society of America (GSA). Boulder, CO, United States. Conference: The Geological Society of America, 96th annual meeting. Indianapolis, IN, United States. October 31-November 3, 1983. Descriptors: Africa; convection; granulites; igneous rocks; kimberlite; Liberia; melting; metamorphic rocks; metasomatism; mineral composition; oxides; petrology; plutonic rocks; silicates; ultramafics; West Africa; igneous and metamorphic petrology. ISSN: 0016-7592.

Haggerty, Stephen E. 1982. "The Mineralogy of Global Magnetic Anomalies. Progress Report, February - August 1982." Performer: Massachusetts Univ., Amherst. Dept. of Geology and Geography. Funded by: National Aeronautics and Space Administration, Washington, DC. 13 August 1982. 47p. Notes: Erts. Report number: NAS126169507; E8310034; NASACR169507; Contract number: NAS526414. Descriptors: *Earth crust; *Liberia; *Magnetic anomalies; *Mineralogy; *Rocks; *South America; Curie temperature; Earth resources program; magnetometers; Magsat satellites; earth sciences and oceanography; geology and mineralogy; NASA earth resources survey program. Abstract: The Curie Balance was brought to operational stage and is producing data of a preliminary nature. Substantial problems experienced in the assembly and initial operation of the instrument were, for the most part, rectified, but certain problems still exist. Relationships between the geology and the gravity and MAGSAT anomalies of West Africa are reexamined in the context of a partial reconstruction of Gondwanaland. NTIS Number: N83135285XSP.

Haggerty, Stephen E. and Toft, Paul B. 1985. "Native iron in the continental lower crust; petrological and geophysical implications." Science. 229; 4714, Pages 647-649. 1985. American Association for the Advancement of Science. Washington, DC, United States. Abstract: information on the mineralogy of the deep continental crust is extremely limited, and the redox state of the lower crust has never been fully addressed. Although the earth's core is probably dominated by metallic iron, terrestrial conditions at the surface are generally oxidized. Naturally occurring iron metal, native iron, is rarely formed except in coal beds, low-temperature (<500|C) serpentinites, lavas that have or may have incorporated carbonaceous sediments, and, at one locality, in a quartz garnet glaucophane lawsonite assemblage. We report the discovery of native iron in lower crustal granulites from Liberia in western Africa. The rocks (2 to 25 cm in diameter) were recovered from a diamond-bearing kimberlite pipe (10|41W, 7|33N) that erupted within the southern shield of the West African cration in the mid-Cretaceous, 90 to 120 million years ago after the breakup of Gondwanaland (4). Granulites and other xenoliths were incorporated from the walls of the volcanic conduit into the kimberlite on its upward passage from the mantle. Although high temperatures were attained, rapid adiabatic cooling has prevented thermal metamorphism of the xenoliths or chemical interaction with the kimberlite. Descriptors: Africa; continental crust; crust; geologic-barometry; geologic-thermometry; granulites; igneous-rocks; inclusions; intrusions; iron; kimberlite; Liberia; lower-crust; magnetic-anomalies; magnetic-properties; metals; metamorphic-rocks; mineralogy; minerals; native-elements; P-T-conditions; pipes; plutonic-rocks; specific-gravity; ultramafics; West Africa; xenoliths; Igneous-and-metamorphic-petrology; mineralogy-of-non-silicates. Map coordinates: LAT: N073300; N070330; LONG: W0104100; W0104100. References: 32; illustrations, including 1 table. ISSN: 0036-8075.

Haggerty, Stephen E. and Tompkins, Linda A. 1983. "Redox state of Earth's upper mantle from kimberlitic ilmenites." Nature (London). 303; 5915, Pages 295-300. 1983. Macmillan Journals. London, United Kingdom. Descriptors: Africa; Antoschka; Fugacite oxygene; fugacity; geochemistry; Guinea; igneous rocks; ilmenite; kimberlite; Koidu; Liberia; mantle; nodules; oxidation; oxides; petrology; plutonic rocks; Potentiel redox; properties; reduction; Sierra Leone; thermodynamic properties; ultramafics; upper mantle; West Africa. References: 54. ISSN: 0028-0836.

Haggerty, Stephen E. and Tompkins, Linda A. 1982. "Opaque mineralogy and chemistry of ilmenite nodules in West Africa kimberlites; subsolidus equilibration and controls on crystallization trends." In: Third international kimberlite conference; abstracts. Terra Cognita. 2; 3, Pages 224-225. 1982. European Union of Geosciences. Strasbourg, France. Conference: Third international kimberlite conference. Clermont-Ferrand, France. 1982. Descriptors: Africa; crystallization; geochemistry; Guinea; igneous-rocks; ilmenite; inclusions; kimberlite; Liberia; low-pressure; Mali; minerals; opaque-minerals; oxides; petrology; plutonic rocks; pressure; Sierra Leone; ultramafics; West Africa; xenoliths; Geochemistry-of-rocks,-soils,-and-sediments; Igneous-and-metamorphic-petrology. ISSN: 0290-9944.

Hall, Chris M.; York, Derek; Onstott, Tullis C. and Hargraves, Robert B. 1984. "40Ar 39Ar spectra models; new techniques for unraveling Precambrian thermal histories." In: GAC-MAC, 1984; program with abstracts; joint annual meeting- AGC, AMC, programme et resumes; reunion annuelle conjointe. Program with Abstracts - Geological Association of Canada; Mineralogical Association of Canada; Canadian Geophysical Union, Joint Annual Meeting. 9; Pages 70. 1984. Geological Association of Canada. Waterloo, ON, Canada. Conference: Geological Association of Canada; Mineralogical Association of Canada; joint annual meeting. London, ON, Canada. May 14-16, 1984. Descriptors: absolute age; Africa; age; alkali feldspar; amphibolites; Ar-Ar; biotite; dates; Encrucijada quartz monzonite; feldspar group; framework silicates; geochronology; granites; Harper amphibolite; igneous rocks; K-feldspar; Liberia; metamorphic rocks; metamorphism; mica group; petrology; plagioclase; plutonic rocks; Precambrian; quartz monzonite; sheet silicates; silicates; South America; Venezuela; West Africa; Geochronology. ISSN: 0701-8738.

Hancox, P. J and Brandt, D. 2000. "An overview of the heavy mineral potential of Liberia." Journal of the South African Institute of Mining and Metallurgy. 100; 1, Pages 29-34. 2000. Abstract: Heavy mineral deposits have been known from Liberia since the 1950s from both river and beach placers. The heavy mineral content of beach sands ranges between 28-62% and the suite includes ilmenite, rutile, zircon and magnetite. These occur together with kyanite, sillimanite, staurolite and garnet. The average ilmenite content is 82% with minor zircon (11%), rutile (6%) and monazite (1%). The ilmenite is, however, of poor quality with haematite intergrowths and a titanium content ranging from 10-42.4% (average+ or -25% TiO (sub 2)). The amounts of rutile and zircon, while of suitable quality, have also previously been deemed too low for economic exploitation. Descriptors: Africa; beach placers; fluvial features; garnet group; heavy mineral deposits; ilmenite; kyanite; Liberia; magnetite; mineral resources; nesosilicates; ore bodies; ore grade; orthosilicates; oxides; placers; potential deposits; rivers; rutile; silicates; sillimanite; staurolite; West Africa; zircon. Notes: SAIMM conference, Heavy minerals 1999, November 15-17, 1999. References: 23; illustrations, including sects., 1 table, geological sketch maps. ISSN: 0038-223X.

Hanson, E. P. 1947. "An Economic Survey of the Western Province of Liberia." Geographical Review. January 1947, Volume 37, pages 53-69. Maps and photos. ISSN: 0016-7428.

Hardman, Mountford N. J. and McGlade, J. M. 2003. "Seasonal and interannual variability of oceanographic processes in the Gulf of Guinea: an investigation using AVHRR sea surface temperature data." International Journal of Remote Sensing, Volume 24, Number 16, August 20, 2003, pages 3247-3268(22). Abstract: The Gulf of Guinea is situated in a critical position for understanding Atlantic equatorial dynamics. This study investigates seasonal and interannual variability in sea surface temperature (SST) throughout this region, focusing on dynamical ocean processes. A 10.5-year time series of remotely sensed SST data with 4 km spatial resolution from the Advanced Very High Resolution Radiometer (AVHRR) were used for this investigation, as they are sufficient to resolve shelf processes. Firstly, patterns of cloud cover were assessed, then spatio-temporal variability in SST patterns was investigated. Features identified in climatological SST images were the Senegalese upwelling influence, coastal upwelling, tropical surface water, river run-off and fronts. Of particular interest is a shelf-edge cooling along the coast of Liberia and Sierra Leone in February. Interannual variability, assessed using annual mean images, time series decomposition and spectral analysis, showed a quasi-cyclic pattern of warm and cool years, perhaps related to El Niño-type forcing. The results of this study show the usefulness of infrared remote sensing for tropical oceanography, despite high levels of cloud cover and atmospheric water vapour contamination, and they provide evidence for theories of westward movement of the upwelling against the Guinea current and remote forcing of the upwelling.

Hargraves, R. B. and Onstott, T. C. 1982. "1.1 Ga rotation of the Kalahari Shield, paleomagnetic evidence from the Guyana and West African Precambrian shields." In: American Geophysical Union; 1982 Spring meeting; abstracts. Eos, Transactions, American Geophysical Union. 63; 18, Pages 309-310. 1982. American Geophysical Union. Washington, DC. Conference: American Geophysical Union; 1982 Spring meeting. Philadelphia, PA, United States. May 31-June 4, 1982. Descriptors: Africa; Bolivar, Venezuela; Bushveld Complex; continental drift; correlation; demagnetization; Free State South Africa; Guyana Shield; Imataca Complex; Irumide mobile belt; Kaapvaal Craton; Kalahari Shield; La Encrucijada Granite; Liberia; Limpopo Belt; magnetization; Namaqualand mobile belt; Orange Free State South Africa; paleomagnetism; plate rotation; plate tectonics; pole positions; Precambrian; remanent magnetization; South Africa; South America; Southern-Africa; tectonophysics; Transvaal-South-Africa; Ubendian-Belt; Venezuela; Vredefort Dome; West Africa; West African Shield; Zimbabwe. ISSN: 0096-3941.

Harley, George Way. 1939. "Roads and Trails in Liberia." Geographical Review, volume 29, July 1939, pages 447-460. Abstract: Technical article describing physiographic regions, trails and trade routes. Dr. Harley's mission station in at Ganta in the interior. ISSN: 0016-7428.

Harley, George Way. 1938. "Liberia." Cambridge, Mass: The Institute. Prepared by the Institute of Geographical Exploration, Harvard University, by G. W. Harley. 1 map, color, 52 x 107cm. Scale 1:534,000. Notes: Relief shown by hachures. Includes notes and inset showing physiographic regions. The data for the region around Cape Mount and the Western Corner was collected by Dr. Junge of the Episcopal Mission at Cape Mount, using pedometer and pocket compass. For North-west area after Captain Henry W. Dennis of the Liberian Frontier forces, whose military sketch map is undoubtedly the best produced by a Liberian. It was compiled and sketched by Henry A.

Kemokai. For Central Liberia from original prismatic compass readings and some scale estimates, checked on main trails by tachometer readings. For the Central Belt from the War College map. For the Saint Paul River Vallet- partly after the preliminary base map of L. C. Daves, done with sketching board and tied into triangulation points. For the area behind New Cess, from a rough map by Rev. G. D. Mellish. This map was tied into the triangulation points and boundary survey made by Ben F. Powell, and has some claim to accuracy. A few details were used from the map by John L. Morris and Josia Massaquoi. The location of the Firestone Plantations and local details are from maps prepared by the Firestone Plantations Company. Other details were obtained by Rev. D. Heydorn, Rev. K. Noltze, Harold R. Bare and George Schwab. The South-eastern portion is most unsatisfactory, being based on the roughest possible estimate. The whole is only roughly sketched, and its only value lies in the fact that it awaits more accurate work. LCCN: 92-684748.

Harvard African Expedition (1926-1927). 1930. The African Republic of Liberia and the Belgian Congo: based on the observations made and material collected during the Harvard African expedition, 1926-1927. Edited by Richard P. Strong. Cambridge: Harvard University Press, 1930. Description: Book. 2 volumes (xxvi, 1064 p.): ill., maps, plates; 27 cm. Series: Contributions from the Department of Tropical Medicine and the Institute for Tropical Biology and Medicine ; no. 5 Contents Notes: v. 1. The African republic of Liberia. Medical and pathological investigations in Liberia and the Belgian Congo. Medical and biological investigations. v. 2. Mammals. Birds. Herpetology. Entomology. Photography. Notes: Includes bibliographical references and index. Subjects: Medical geography Liberia. Medical geography Congo (Democratic Republic). Natural history Liberia. Natural history Congo (Democratic Republic). Zoology Liberia. Zoology Congo (Democratic Republic). Botany Liberia. Tropical medicine Liberia. Tropical medicine Congo (Democratic Republic). Scientific expeditions Africa. Liberia Description and travel. Liberia Social life and customs. Other: Strong, Richard Pearson, 1872-1948. Held at the USGS Library- Call number: 590(756) H26a. LCCN: 30-28936.

Harvard University. Institute of Geographical Exploration. Liberia. Topographic map: 1:534,000. np. 1938. 1 sheet in color, mountains shown by "wooly caterpillars." Printed in Great Britain. Held in the USGS Library.

Hasselmann, Karl-Heinz. 1991. "Les Cahiers d'outre-mer." Translated title: "Books of Overseas." Bordeaux, 1991. Descriptors: Forests- Liberia. Notes: English summary.

Hasselmann, Karl-Heinz. 1979. Liberia, geographical mosaics of the land and the people. Monrovia: Ministry of Information, Cultural Affairs & Tourism. 278 pages: illustrations; 26 cm. Descriptors: Physical geography- Liberia; Liberia- Economic conditions; Liberia- Population. Notes: Bibliography: pages 271-278. LCCN: 80-120192; OCLC: 6920329.

Hasselmann, Karl-Heinz. 1977. "Geography of Greenville." Monrovia, Liberia: Dept. of Geography, University of Liberia. 156 pages: maps; 29 cm. Series: Occasional research paper, Dept. of Geography, University of Liberia; no. 17; Variation: University of Liberia. Dept. of Geography. Occasional research paper - Dept. of Geography, University of Liberia; no. 17. Descriptors: Greenville (Liberia)- Description and travel; Greenville (Liberia)- History; Greenville (Liberia)- Social conditions- Statistics. Notes: "July 26, 1977." OCLC: 38636547.

Hasselmann, Karl-Heinz. 1977. Liberia: Population, Size, and Density, 1962 and 1974. Monrovia: Dept. of Geography, University of Liberia. 62 pages: illustrations, maps; 21 cm. Occasional research paper, Department of Geography, University of Liberia; no. 16; Variation: University of Liberia. Dept. of Geography. Occasional research paper, Dept. of Geography, University of Liberia; no. 16. Descriptors: Liberia- Population; Liberia-Administrative and political divisions. Note(s): Includes bibliographical references (p. 60-62). OCLC: 6072095.

Hasselmann, Karl-Heinz. 1976. Centrifugal and centripetal population variation in Liberia. Monrovia: Dept. of Geography, University of Liberia. 1 map; photocopy; 51 x 49 cm. Scale [ca. 1:1,000,000]. Descriptors: Liberia- Population- Maps. Notes: Shows growth between 1962 and 1974. "August 1976." Calculations, cartography, and drawing by Karl-Heinz Hasselmann. LCCN: 85-690730; OCLC: 12942371.

Hasselmann, Karl-Heinz. 1975. "Gbarnga." Monrovia: University of Liberia, Dept. of Geography. 1 map; photocopy; 70 x 85 cm. Scale 1:2,500. Subjects: Gbarnga (Liberia)- Maps. Note(s): "July 26, 1975." Includes index to points of interest. LCCN: 85-690736; OCLC: 12836361.

Hasselmann, Karl-Heinz. 1975. "Gbarnga, Liberia: an economic-geographic survey." Monrovia: Dept. of Geography, University of Liberia. 45 leaves; 28 cm. Occasional research paper, Department of Geography, University of Liberia; no. 8; Variation: University of Liberia, Dept. of Geography; Occasional research paper, Department of Geography, University of Liberia; no. 8. Descriptors: Gbarnga (Liberia)- Economic conditions. Notes: Photocopy. Bibliography: leaf 45. LCCN: 78-326959; OCLC: 2778118.

Hasselmann, Karl-Heinz. 1974. "Map of Zwedru, Tchien." [Monrovia]: Dept. of Geography, University of Liberia. 1 map; photocopy; 63 x 98 cm. Scale [ca. 1:2,500]. Descriptors: Zwedru (Liberia)- Maps. Note(s): Shows points of interest. "Aerial photography by Hansa Luftbild, 1972." Oriented with north toward the lower left. "Mapping, airphotos, outline, and drawing, May to July 1974, by Dr. K.H. Hasselmann, Department of Geography, University of Liberia (Fac. of Liberal & Fine Arts)." LCCN: 85-690734; OCLC: 12942420.

Hasselmann, Karl-Heinz. 1974. "Robertsport." Monrovia: University of Liberia, Dept. of Geography. 1 map; photocopy; 81 x 81 cm. Scale 1:2,500. Descriptors: Robertsport (Liberia)- Maps. Notes: "Base: Hansa Luftbild, aerial photography, 1972." Oriented with north toward the lower left. Includes index to points of interest. Mapping and drawing by Karl-Heinz Hasselmann, February 1974. LCCN: 85-690729; OCLC: 12942370.

Hasselmann, Karl-Heinz and Kaye, James. 1976. "Sanniquellie." Monrovia: University of Liberia, Dept. of Geography. 1 map; photocopy; 70 x 67 cm. Scale not given. Descriptors: Sanniquellie (Liberia)- Maps. Notes: Includes index to points of interest. LCCN: 85-690738; OCLC: 12836534.

Hasselmann, Karl- Heinz and Srivastava, M. L. 1976. "Liberia Population Development Map, 1962-1974: Based on Clans and Townships." Monrovia: Department of Geography, University of Liberia. 1 map, photocopy, 98 x 90 cm. Scale 1:550,000. Notes: "August 1976/2." LCCN: 85-690739.

Hastings, David A. 1983. "An updated Bouguer anomaly map of south-central West Africa." Geophysics. 48; 8, Pages 1120-1128. 1983. Society of Exploration Geophysicists. Tulsa, OK, United States. Abstract: A new Bouguer gravity anomaly map compiled for western Africa adds data for Ghana, Guinea, and Liberia. The new data add detail to a key part of the Eburnean shield and assist in the development of a model of rifting at the time of the Eburnean orogeny, 2000 million years ago. This model includes a framework for the deposition of the region's mineral deposits. The model and existing field data can be used to guide future minerals exploration in the region. Descriptors: Africa; aluminum ores; Benin; Bouguer anomalies; Burkina Faso; diamonds; evolution; geophysical methods; geophysical surveys; gold ores; gravity anomalies; gravity methods; gravity survey maps; Guinea; iron ores; Ivory Coast; Liberia; Mali; manganese ores; maps; Mauritania; metal ores; mineral deposits, genesis; mineral exploration; Niger; orogeny; Pan African Orogeny; plate tectonics; Precambrian; processes; Proterozoic; rifting; surveys; Togo; upper Precambrian; West Africa; Applied geophysics. References: 55; illustrations, including sketch maps. ISSN: 0016-8033.

Hastings, David A. 1981. "An interpretation of the preliminary total-field Magsat anomaly map." In: The Geological Society of America, 94th annual meeting. Abstracts with Programs - Geological Society of America. 13; 7, Pages 469. 1981. Geological Society of America, Boulder, CO, United States. Conference: Geological Society of America, 94th annual meeting; the Paleontological Society (73rd); the Mineralogical Society of America (62nd); the Society of Economic Geologists (61st); Cushman Foundation (32nd); Geochemical Society (26th); National Association of Geology Teachers (22nd); Geoscience Information Society (16th). Cincinnati, OH. November 2-5, 1981. Descriptors: Africa; African Shield; anomalies; Atlas Mountains; cartography; Central Africa; Central African Republic; geophysical methods; geophysical surveys; global; interpretation; Liberia; Magsat; Nigeria; North Africa; Precambrian; remote sensing; satellite methods; surveys; uplifts; West Africa. ISSN: 0016-7592.

Hastings, David A. 1979. "An updated Bouguer anomaly map of South-central West Africa." Abstracts - Society of Exploration Geophysicists International Meeting. 49, Pages 65. 1979. Society of Exploration Geophysicists, International Meeting and Exposition. Tulsa, OK, United States. Conference: Society of Exploration Geophysicists, 49th annual international meeting. New Orleans, LA., United States. November 4-8, 1979. Descriptors: Africa; Bouguer anomalies; diamonds; economic geology; geophysical methods; geophysical surveys; Ghana; gold ores; gravity anomalies; gravity methods; gravity survey-maps; Guinea; iron ores; Liberia; manganese ores; maps; metal ores; mineral deposits, genesis; mineral exploration; mineral resources; ore deposits; plate tectonics; processes; rifting; structural controls; surveys; suture zones; tectonophysics; West Africa; Solid earth geophysics. ISSN: 0740-543X.

Hazelwood, P. T. 1980. "Environmental Profile of Liberia, Phase I." Performer: Library of Congress, Washington, DC. Science and Technology Project. Sponsor: Agency for International Development, Washington, DC. July 1980. 72 pages. Report: AIDPN-AAG-978. Sponsored in part by the National Committee for Man and the Biosphere, Washington, DC. Descriptors: Natural resources; Liberia; Environmental surveys; water resources; wildlife; forestry; agriculture; mineral deposits; economic factors; soils; population growth; pollution. Abstract: Liberia has a tropical climate favoring high forest vegetation and is rich in natural resources, notably iron ore, timber, and rubber. Its major topographical and bio-geographical regions include a coastal plain, a belt of rolling hills, a belt of low mountain ranges and plateaus, and northern highlands. This report is a preliminary review of information available in the United States on Liberia's environment and natural resources. Topics covered by the report include the nation's physical, demographic, and social and economic characteristics; renewable resources; non-renewable resources; parks, reserves, and other protected areas. NTIS Report Number: PB82130667.

Heatherington, A. L; Mueller, P. A and Dallmeyer, R. D. 1993. "Geochemical provenance of Florida basement components." In: Geological Society of America, Southeastern Section, 42nd annual meeting. Abstracts with Programs - Geological Society of America. 25; 4, Pages 22-23. 1993. Geological Society of America (GSA). Boulder, CO, United States. 1993. Conference: Geological Society of America, Southeastern Section, 42nd annual meeting. Tallahassee, FL, United States. April 1-2, 1993. Descriptors: absolute-age; Africa; Archean; basement; Birrimian; block structures; clastic rocks; cooling; correlation; dates; exotic terranes; Florida; Gondwana; Grenvillian Orogeny; Liberia; northern South America; Osceola County, Florida; paleomagnetism; Paleoproterozoic; Paleozoic; Precambrian; Proterozoic; provenance; Rb-Sr; rifting; sandstone; sedimentary rocks; Sm-Nd; South America; Suwannee Basin; terranes; U-Pb; United States; upper Precambrian; West Africa; Structural geology; Geochronology. ISSN: 0016-7592.

Hedge, C. E.; Marvin, R. F. and Naeser, C. W. "Analytic Age provinces in the basement rocks of Liberia." Journal of Research. Volume: 3. Issue Number: 4. Pages (analytic): 425-429. August 1, 1975. U. S. Geological Survey, Reston, VA. Subjects: absolute age; Africa; basement; data; dates; geochronology; K/Ar; Liberia; metamorphic rocks; Precambrian; provinces; Rb/Sr; USGS; West Africa. Illustrations, including tables, sketch map. ISSN: 0091-374X.

Hempton, Mark. 1995. "Opening history of the Mesozoic Equatorial Atlantic continental margins." In: American Association of Petroleum Geologists 1995 annual convention. Annual Meeting Abstracts - American Association of Petroleum Geologists and Society of Economic Paleontologists and Mineralogists. 4; Pages 41. 1995. American Association of Petroleum Geologists and Society of Economic Paleontologists and Mineralogists. Tulsa, OK, United States. 1995. American Association of Petroleum Geologists 1995 annual convention. Houston, TX, United States. March 5-8, 1995. Descriptors: Africa; Atlantic-Ocean; Benue-Valley; Brazil; clastic-rocks; crust; data-processing; Demerara-Plateau; dike-swarms; Equatorial-Atlantic; evolution; extension-tectonics; faults; Guinea-Plateau; intrusions; Ivory Coast; Liberia; Marajo-Basin; marine-environment; Mesozoic; oceanic-crust; rates; reconstruction; sea-floor-spreading; sedimentary rocks; sedimentation; shale; South-America; strain; strike-slip-faults; subsidence; tectonics; volcanism; West Africa; wrench-faults. ISSN: 0094-0038.

Hempton, Mark R. 1993. "Regional paleogeographic evolution of West Africa; implications for hydrocarbon exploration." In: AAPG distinguished lecture tours, 1993-1994. AAPG Bulletin. 77; 11, Pages 2020. 1993. American Association of Petroleum Geologists. Tulsa, OK, United States. 1993. Descriptors: Africa; Angola; Benin; Benue Valley; Cameroon; Central Africa; continental margin; data processing; deltaic environment; evolution; faults; geometry; kinematics; lacustrine environment; Liberia; Mauritania; Namibia; Niger Delta; Nigeria; paleoclimatology; paleogeography; petroleum exploration; reconstruction; rift zones; shear; Sierra Leone; source rocks; South Africa; Southern Africa; turbidite; West Africa; wrench faults; economic geology, geology of energy sources; solid earth geophysics. ISSN: 0149-1423.

Hermann, Chris; Shaw, Margaret and Hannah, John. 1985. Development Management in Africa: The Case of the Agriculture Analysis and Planning Project in Liberia. Special study. Agency for International Development, Washington, DC. Center for Development Information and Evaluation. Performer: Development Alternatives, Inc., Washington, DC. December 1985. 56p. Notes: Prepared in cooperation with Development Alternatives, Inc., Washington, DC. Descriptors: Africa; developing countries; data acquisition; data processing; management methods; specialized training; national government; budgeting; agriculture; *project planning; *Liberia. developing country application; agricultural sector; sector planning; technical assistance; agriculture-and-food-agricultural economics; administration and management; management practice. Abstract: The central objective of the Agriculture Analysis and Planning Project and its predecessor, the Agriculture Development Program, was to improve the capability of the Government of Liberia's Ministry of Agriculture for the data collection and analysis necessary for sector planning. Key lessons learned are: sustainable data-related technologies must be simple and low-cost, but sufficient for basic information requirements; the transfer of technology creates management demands which must be anticipated and used as criteria in selecting technical assistance; technical assistance and training must be continuous to assure that gains made are not lost; in-country training must be emphasized to minimize the disruption of agency operations; necessary organizational changes must be anticipated; advisors should provide quality control and staff support if the agency does not; project planning should be flexible. NTIS Number: PB87155115XSP; OCLC: 17247035.

Heseltine, N. 1973. "Mining And Agriculture. Development Or Disappearance? The Case Of Liberia." World Crops. Volume 25, No. 3 (May/June 1973), pages 128-131. Map. ISSN: 0043-8391; OCLC: 1770135.

Hildebrand, Robert S. 1994. "Are kimberlites lower plate magmatism triggered by plate collisions?" In: Geological Society of America, 1994 annual meeting. Abstracts with Programs - Geological Society of America. 26; 7, Pages 312. 1994. Geological Society of America (GSA). Boulder, CO, United States. 1994. Conference: Geological Society of America 1994 annual meeting. Seattle, WA, United States. October 24-27, 1994. Descriptors: Africa; Appalachians; Archean; basins; Canada; Cenozoic; Colorado; continental crust; convection; Cretaceous; crust; diamonds; Eastern Canada; emplacement; Eocene; flexure; fore arc basins; gems; igneous activity; igneous rocks; Innuitian Orogeny; Ivory Coast; kimberlite; Liberia; magmas; Malay Archipelago; mechanism; Mesozoic; models; New Guinea; North America; North American Cordillera; Northwest Territories; Paleogene; petrography; plate collision; plate tectonics; plutonic rocks; Precambrian; Senegal; South Africa; Southern Africa; Southern U.S.; Tertiary; ultramafics; United States; West Africa; Western Canada. ISSN: 0016-7592.

Hill, L. J. 1987. "Modeling the macroeconomy/energy economy relationship in developing countries: the case of Liberia." Journal of Developing Areas. Volume 22, no.1 (1987) pages 71-84. Abstract: It is possible to construct a relatively simple modeling system that captures the major interactions between international economic activity, the domestic economy, and the energy sector of developing economies. This paper discusses the construction of one such system which was used to simulate energy demand by sector and fuel type in Liberia, West Africa, over the 1982-2000 time period. The paper is divided into four sections. The second section provides an overview of the Liberian economy; its discussion on Liberia's output, export-import structure, and the energy sector serves as a preface for the discussion of the specification of the modeling system presented in the third section. The fourth section discusses the simulation results under four different scenarios. The final section presents some conclusions of the study.

Hill, L. J. 1984. Liberian Macroeconomy and Simulation of Sectoral Energy Demand: 1981-2000. Oak Ridge National Lab., TN, United States. Funded by: Department of Energy, Washington, DC. June 1984. 99p. Contract number: AC0584OR21400; Report number: ORNLTM9065. NTIS Advisory: Portions are illegible in microfiche products. Original copy available until stock is exhausted. Descriptors: charcoal; electric power; fuel oils; gas oils; gasoline; iron ores; jet engine fuels; kerosene; Liberia; natural rubber; petroleum; wood; data compilation; econometrics; energy analysis; energy demand; exports; forecasting; imports; prices; simulation. economic analyses; energy utilization; behavioral and social sciences economics; energy conversion; non propulsive conversion techniques; energy use supply and demand; energy policies regulations and studies; business and economics; foreign industry development and economics. Abstract: The primary purpose of this report is to document the results of a research effort on end-use, sector energy demand in Liberia, West Africa over the 1981-2000 time horizon. The research was undertaken as one component of a much broader integrated energy assessment of Liberia. Other components of the assessment, however, focused on current energy supply and consumption together with future energy supply options for Liberia. This particular report is devoted exclusively to a discussion of Liberian energy demand. The methodology utilized to simulate Liberian sectoral energy demand over the period 1981-2000 involved the recursive interaction of a macroeconomic model and individual, econometrically-estimated sectoral demand equations. That is, given the projections for gross output in the Liberian economy from the macroeconomic model, sectoral energy demand was simulated. The individual energy demand equations were estimated on the basis of economic variables that are theorized to influence energy consumption in the respective sectors (e.g., price, output). The primary conclusion drawn from the analysis is that, besides being sensitive to changes in international economic activity, the demand for energy in Liberia over the 1981 to 2000 horizon is highly sensitive to internal production of its two primary exports: iron ore and rubber. More specifically, as characterized in the four scenarios, future growth in Liberian energy demand is contingent on the output of three companies: the Liberian American Swedish Mining Company, the Bong Mining Company, and the Firestone Rubber Company. Therefore, expansion of Liberia's energy supply capacity in the future should proceed cautiously. 16 references, 6 figures, 15 tables. NTIS Number: DE84013665XSP.

Höll, C. and Kemle, von Mücke S. 2000. " Late Quaternary Upwelling Variations in the Eastern Equatorial Atlantic Ocean as Inferred from Dinoflagellate Cysts, Planktonic Foraminifera, and Organic Carbon Content. Quaternary Research, Volume 54, Number 1, July 2000, pp. 58-67(10). Abstract: Analysis of multiple proxies shows that eastern equatorial Atlantic upwelling was subdued during isotope stage 5.5, more intense during stages 4, 5.2, 5.4, and 6, and most intense early in stage 2. These findings are based on proxy measures from a core site about 600 km southwest of Liberia. The proxies include total organic carbon content, the ratio of peridinoid and oceanic organic walled dinoflagellate cyst species, accumulation rates of calcareous dinoflagellates, estimates of sea surface paleotemperatures, the difference in stable oxygen isotope composition between two species of planktonic foraminifera that live at different water depths, and the abundance of the planktonic foraminifera Neogloboquadrina dutertrei. Most of these parameters consistently vary directly or inversely with one another. Slight discrepancies between the individual parameters show the usefulness of a multiple proxy approach to reconstruct paleoenvironments. Our data confirm that northern summer insolation strongly influences upwelling in the eastern equatorial Atlantic Ocean.

Holsoe, Torkel. 1961. Third report on forestry progress in Liberia, 1951-1959. Washington, DC: United States. International Cooperation Administration. Descriptors: Forests and forestry- Liberia. Note(s): Cover title. The two previous reports by Torkel Holsoe are Forestry opportunities in the Republic of Liberia, 1954, and Forestry progress and timbering opportunities in the Republic of Liberia, 1956 [sic]. Includes bibliographical references. OCLC: 399334.

Holsoe, Torkel. 1955. Forestry progress and timbering opportunities in the Republic of Liberia. Washington, DC: United States. International Cooperation Administration. 92 pages, illustrations, map. Descriptors: Forests and forestry- Liberia. LCCN: 56-60372; OCLC: 3653991.

Holsoe, Torkel. 1954. Forestry opportunities in the Republic of Liberia. Washington, DC: United States. Foreign Operations Administration. 41 pages, illustrations. Descriptors: Forests and forestry- Liberia. [by] Torkel Holsoe, forestry adviser. Foreign Agricultural Service and Forest Service, U.S. Dept. of Agriculture, cooperating. LCCN: 54-60288; OCLC: 4670101.

Hottes, Karlheinz. 1971. Liberia. 1971 [neunzehnhunderteinundsiebzig]: Ergebnisse e. Studienbereisung durch e. trop. Entwicklungsland Karlheinz Hottes [u. a.]. Paderborn: Scheoningh, 1973. 166 p.: ill., maps; 29 cm. Bochumer geographische Arbeiten: Heft 15. Notes: Includes bibliographies. Subjects: Liberia- Economic conditions. ISBN: 3506711954; OCLC: 1237598.

Hurley, P. M.; Fairbairn, H. W. and Gaudette, H. E. 1976. "Progress report on early Archean rocks in Liberia, Sierra Leone and Guayana, and their general stratigraphic setting." In: The early history of the Earth. Windley, B. F. (editor). Pages 511-521. 1976. John Wiley & Sons. New York, N. Y., United States. Descriptors: absolute age; Africa; amphibolites; Archean; gneisses; granulites; Guyana; interpretation; Liberia; metamorphic-rocks; metamorphism; petrology; Precambrian; review; Sierra Leone; South America; West Africa; igneous and metamorphic petrology. Illustrations, including sketch map.

Hurley, P. M.; Leo, G. W.; White, R. W. and Fairbairn, H. W. 1971. "Liberian age province (about 2,700 m.y.) and adjacent provinces in Liberia and Sierra Leone." Geological Society of America Bulletin. 82; 12, Pages 3483-3490. 1971. Geological Society of America (GSA). Boulder, CO, United States. Descriptors: absolute age; Africa; Archean; basement; Cambrian; dates; isochrons; Kambui Schist; Kasila Group; Liberia; Marampa Formation; metamorphic rocks; metasedimentary rocks; metavolcanic rocks; Nimba Mountains; Paleozoic; Pan African Orogeny; Precambrian; Proterozoic; Rb-Sr; Sierra Leone; upper Precambrian; West Africa; whole rock. References: 22; illustrations, including sketch map. ISSN: 0016-7606.

Hurley, P. M.; Leo, G. W.; White, R. W. and Fairbairn, H. W. 1970. "The Liberian age province (ca. 2700 m.y.), and adjacent provinces in Liberia and Sierra Leone." Abstracts with Programs - Geological Society of America. 2; 7, Pages 583. 1970. Geological Society of America (GSA).

Boulder, CO, United States. Descriptors: absolute age; Africa; dates; Liberia; metamorphic rocks; Precambrian; Sierra Leone; West Africa; Geochronology. ISSN: 0016-7592.

Hütten, Horst G. 1977. "Impressions of Liberia: from the sketch-books of a journey to the Bong Range Iron Ore Mine in Liberia." Translated title: "Mit dem Zeichenstift Afrika: aus den Skizzenbüchern einer Reise zu der Eisenerzmine Bong Mine in Liberia." Mettmann, Markt 17: H.G. Hütten [Selbstver.]. 72 pages: chiefly ill.; 27 cm. Language: English. Descriptors: Bong Mining Company; Liberia- Description and travel. Notes: English and German. Other Titles: Mit dem Zeichenstift in Afrika. ISBN: 3921986125; LCCN: 81-137160; OCLC: 9919282.

Huvane, Kathleen. 2001. "U.N. cracks down on "conflict diamonds" (The diamond trade in Liberia) (Brief Article)." World Watch. 14. September 2001: page 9. Abstract: In an attempt to stem international trade in "conflict diamonds" from West Africa, on May 7, 2001 the United Nations imposed economic sanctions on Liberia, which has smuggled a steady flow of ill-gotten diamonds from rebel forces in neighboring Sierra Leone... ...Liberia's forest industry has played an integral role in propagating the diamond-weapons trade. In addition to providing corrupt government officials and businessmen with revenue through the sale of illegally harvested tropical timber, the roads and vehicles used to transport felled trees have facilitated the movement of weapons and diamonds. Since the timber processing industry in Liberia is virtually nonexistent, the vast majority of logs are exported whole, further depriving locals of employment opportunities they might reap from their forests. Official Forestry Department statistics for the first six months of 2000, which do not account for the vast illegal timber harvest, reveal that logging during that time period exceeded that of the previous four years combined. Not counting illegal logging, the forestry department projects that Liberian forests may be commercially viable for only another ten years. Halting the illegal trade will be difficult in the face of the region's fighting and corruption. For instance, Dutch national Gus Koewenhoven, the notorious "Godfather of Liberia," heads two logging companies with significant concessions in Liberia and also serves as a board member of the Liberian Forestry Development Association (FDA)- the government agency charged with protecting and monitoring Liberia's forests. Koewenhoven has used his considerable connections to carve trade routes that allow for easy collection and transfer of timber, diamonds, and arms. International pressure is also keeping this illegal trade flowing: China and France, the primary importers of Liberian timber, used their political muscle to ensure that the economic sanctions imposed in May did not include forest products. In addition, British and Ukrainian companies have supplied arms to Sierra Leone's RUF rebels, commonly funneled through Liberian channels.

"Infrastructural survey report for the development of the Wologisi iron mining in the Republic of Liberia." 1979. Kokusai Kyoryoku Jigyodan. [Tokyo]: Japan International Cooperation Agency. Description: 2 v.: ill. (some color); 30-31 cm. Descriptors: Roads- Liberia; Harbors- Liberia; Iron mines and mining- Liberia- Lofa County. Notes: "MPI CR (I) 79-2; MPI CR (I) 79-31"--Cover. Includes indexes. LCCN: 84-256610; OCLC: 11973611.

"International Trade: Significant Challenges Remain in Deterring Trade in Conflict Diamonds. " GAO, Washington, DC (USA). [vp]. 13 February 2002. Descriptors: crime; money laundering; diamond industry; conflict resolution; arms control; government regulations; international trade; Liberia; Sierra Leone. Abstract: As a high-value commodity easily concealed and transported and virtually untraceable to their original source, diamonds are used in lieu of currency in arms deals, money laundering, and other criminal activity. U.S. controls over diamond imports only require certification from the country of last import--and thus cannot identify diamonds from conflict sources. Although the United States bans diamonds documented as coming from the National Union for the Total Independence of Angola, the Revolutionary United Front in Sierra Leone, and Liberia- all of which are subject to U.N. sanctions- this does not prevent such diamonds shipped to a second country from being mixed into parcels destined for the United States. GAO found that the Kimberley Process's proposal for international diamond certification incorporated some elements of accountability. However, it is not based on a risk assessment, and some high risk activities are subject only to "recommended" controls. Also, from the time when rough diamonds enter the first foreign port until the final point of sale there exists only a voluntary industry participation and self-regulated monitoring and enforcement system. These and other shortcomings significantly undermine efforts to deter trade in questionable diamonds.

"Iron may Dominate Rubber as Liberia's Main Export." 1959. East African Trade and Industry. Volume 6, number 66, pages 37, 39. LCCN: 60-43580; OCLC: 2446498.

Jablonski, Donna M. (editor). 1982. "Mideast and Africa." In the collection: Future energy sources; National development strategies. Volume 1; 1982. McGraw-Hill. Washington, DC. Pages: variously paginated. Descriptors: Africa; Asia; East Africa; economic geology; Egypt; energy sources; exploration; Indian Peninsula; Israel; Jordan; Kenya; Liberia; management; Middle East; Nigeria; North Africa; Pakistan; policy; possibilities; programs; resources; Sierra Leone; Sudan; Turkey; utilization; West Africa. Illustrations, including tables.

Jacobi, W. J. 1916. "Map of Liberia." United States. War Dept. General Staff. War College Division. Washington: The Division. Description: 1 map: mounted on linen; 73 x 62 cm. Descriptors: Tribes- Liberia- Maps; Liberia- Maps. Map Info: Scale 1:750,000; (W 11030'--W 7030'/S 9000'--S 4030'). Notes: Relief shown by form lines. Shows names of tribes, sections, and tribal divisions. Prepared in the War College Division, General Staff; W. J. Jacobi, draftsman. OCLC: 54646827.

Jahns, Susanne; Huels, Matthias and Sarnthein, Michael. 1998. "Vegetation and climate history of West Equatorial Africa based on a marine pollen record off Liberia (site GIK 16776) covering the last 400,000 years." Review of Palaeobotany and Palynology. 102; 3-4, Pages 277-288. 1998. Elsevier. Amsterdam, Netherlands. 1998. Abstract: Based on pollen analysis of a sediment core from the Atlantic Ocean off Liberia the West African vegetation history for the last 400 ka is reconstructed. During the cold oxygen isotope stages 12, 10, 8, 6, 4, 3 and 2 an arid climate is indicated, resulting in a southward shifting of the southern border of the savanna. Late Pleistocene glacial stages were more arid than during the Middle Pleistocene. A persistence of the rain forest in the area, even during the glacial stages, is recorded. This suggests a glacial refuge of rain forest situated in the Guinean mountains. Afromontane forests with Podocarpus occurred in the Guinean mountains from the stages 12 to 2 and disappeared after. The tree expanded from higher to lower elevations twice in the warm oxygen isotope stage 11 (pollen subzones 11d, 11b) and at least twice during the warm stage 5 (pollen subzones 5d, 5a), indicating a relative cool but humid climate for these periods. Descriptors: Africa; arid-environment; Atlantic-Ocean; biostratigraphy; biozones; Cenozoic; cores; East-Atlantic; equatorial region; floral list; forests; GIK-16776-Core; glaciation; Holocene; humid environment; isotope ratios; isotopes; Liberia; marine environment; marine sediments; microfossils; miospores; O-18-O-16; oxygen; paleoclimatology; paleoecology; paleotemperature; palynomorphs; Plantae; Pleistocene; Podocarpus; pollen; pollen-analysis; pollen diagrams; quantitative analysis; Quaternary; rain forests; savannas; sedimentation; sedimentation rates; sediments; stable isotopes; terrestrial environment; tropical environment; vegetation; West Africa; Quaternary-geology. Map Coordinates: Lat: N034401; N034401; Long: W0112309; W0112309. Notes: Includes appendices. References: 62; illustrations, including 3 tables, sketch map. ISSN: 0034-6667.

Janse, A. J. A. 1996. "A History of Diamond Sources in Africa." Gems and Gemology. Volume 31 (1). Spring 1996. Pages 2-30. ISSN: 0016-626X; OCLC: 7921950.

John, D. M. 1976. "The marine algae of Ivory Coast and Cape Palmas in Liberia (Gulf of Guinea)." Revue Algologique. 11; 3-4, Pages 303-324. 1976. Laboratoire de Cryptogamie. Paris, France. Descriptors: Africa; algae; Atlantic Ocean; Cape Palmas; floral list; Gulf of Guinea; Ivory Coast; Liberia; microfossils; North Atlantic; occurrence; paleobotany; Plantae; thallophytes; West Africa; Paleobotany. Illustrations. ISSN: 0035-0702.

Johnson, D. H.; Holmes, A. J. and Cooper, B. R. 1971. "Geochemical investigations of base metal occurrences in western Liberia." In: Geochemical exploration (International Geochemical Exploration Symposium, 3rd, Proc.). Special Volume - Canadian Institute of Mining and Metallurgy. 11; Pages 195. 1971. Canadian Institute of Mining and Metallurgy. Montreal, PQ, Canada. Descriptors: Africa; economic geology; exploration; geochemical methods; geochemical surveys; Liberia; metals; mineral exploration; soils; surveys; West Africa; Economic geology of ore deposits. ISSN: 0576-5447.

Johnson, D. H; Holmes, A. J. and Cooper, B. R. 1970. "Geochemical investigations of base metal occurrences in western Liberia." In: Third international geochemical exploration symposium; program and abstracts. Boyle, R.W. Pages 41-42. 1970. Can. Inst. Mining Met., Geol. Div.-Soc. Econ. Geol.. Toronto, ON, Canada. Conference: Third international geochemical exploration symposium. Toronto, ON, Canada. April 16-18, 1970. Descriptors: Africa; economic geology; Galena creek; Gboeya creek; geochemical methods; Liberia; metals; mineral exploration; Sam Davis creek; soils; West Africa.

Johnson, Donald Haskall and White, Richard W. 1969. "Bibliography of the Geology and Mineral Industry of Liberia and Adjacent Countries." Washington: U.S. Geological Survey; Liberian Geological Survey; United States. Agency for International Development. 65 pages; 28 cm. Series: US Geological Survey Open-file report; 1209. Series: Memorandum report. Liberian Geological Survey; no. 40-A; Variation: Memorandum report (Liberian Geological Survey); no. 40-A. Descriptors: Geology- Liberia- Bibliography; Mines and mineral resources; Mineral industries-Liberia- Bibliography. Notes: "September 1969." "USGS IR-LI-44." Referred to in press release dated April 21, 1969. A bibliography compiled jointly by the Liberian Geological Survey and the U.S. Geological Survey for the Government of Liberia and the U.S. Agency for International Development. "This compilation represents an attempt to list, as nearly as possible, all references pertaining to the geology and mineral industry of Liberia, and to indicate those that are held by the library of the Liberian Geological Survey. The bibliography is a contribution to the Geological Exploration and Resource Appraisal Program, a combined effort of the Government of Liberia and the United States Agency for International Development, carried out jointly by the Liberian Geological Survey and the United States Geological Survey. The references were compiled from many sources, including the card catalog of the US Geological Survey Library in Washington, the holdings of the Liberian Geological Survey library, various unpublished bibliographies, references cited by authors of publications dealing with Liberian geology, and chance encounters in the literature. The list is admittedly incomplete… References to mining, metallurgy and mineral economics involving Liberia have been included, as have many papers dealing with physical geography, cartography, pedology, hydrology and related subjects." LCCN: 90-101134; OCLC: 22180090.

Johnson, Donald Haskall; Holmes, A. J. and Cooper, B. R. 1968. "Geochemical Investigations of Base Metal Occurrences in Western Liberia." Paper presented at the September monthly meeting of the Liberian Geological Society in 1968. Published in Annual Publication, pages 78-79.

Johnson, Jahmale C. and Gentry, Donald W. 1977. An evaluation of mine taxation in Liberia and its effects on investment decisions. Description: xi, 146 leaves: ill.; 29 cm. Descriptors: Mines and mineral resources; Liberia-Taxation; Mineral industries; Finance; Mineral industries; Taxation. Notes: Typescript (photocopy). Includes bibliographical references (leaves 141-146). Master's Thesis (M. Sc.), Colorado School of Mines, 1977. OCLC: 28537961.

Johnson, T. 1967. Mano River mine report. Bulletin of the Geological, Mining and Metallurgical Society of Liberia. 2; Pages 36-47. 1967. Geological, Mining and Metallurgical Society. Monrovia, Liberia. Descriptors: iron deposits; prospecting and exploration; mineralized zones; Liberia; Africa; economic geology; Grand Cape Mount County, iron; Mano River; metals; mine report; West Africa; Economic geology of ore deposits. Illustrations, including sketch maps. ISSN: 0367-4819.

Johnson, T. 1966. "Mano River Mine Report." Paper presented at the December monthly meeting of the Liberian Geological Society in 1966. Published in Annual Publication, pages 36-48.

Johnston, Harry Hamilton, Sir, 1858-1927. and Stapf, Otto; b. 1857. 1906. Liberia. Schomburg Collection of Negro Literature and History. London: Hutchinson. Description: 2 volumes: illustrations, portraits, maps. Contents: Volume I. Liberia. Ancient history. Normans and Genoese. Portuguese. Pepper and gold. The Guinea trade in the sixteenth and seventeenth centuries. A Dutch account of Liberia in the seventeenth century. The slave trade. The founding of Liberia. The last phase of the slave trade. Governors of Liberia. Independence. President Roberts. Frontier questions. The loan and its consequences. Recent history. The Americo-Liberians. Commerce. Geography of Liberia. Climate and rainfall. Geology and minerals. Volume II. Flora of Liberia. Fauna. Anthropology. Folklore. Languages: The languages of Liberia. Appendix IX: The Vai syllabarium or alphabet. Vocabularies of Liberian and other West African languages. Index. Descriptors: Natural history- Liberia; Botany- Liberia. Vai language. Geographic: Liberia. Africa, West- Languages. Notes: Paged continuously; v. 1: xxviii, 519, [1] p.; v. 2: xvi, 521-1183 pages. Microfilmed from the Schomburg collection of the New York Public Library. Includes bibliographical references (v. 1, xiii-xvii. 1971. 1 microfilm reel. 35 mm. By Sir Harry Johnston ... with an appendix on the flora of Liberia, by Dr. Otto Stapf; 28 coloured illustrations, by Sir Harry Johnston, 24 botanical drawings by Miss Matilda Smith, 402 black and white illustrations from the author's drawings and from photographs by the author and others, 22 maps by Mr. J. W. Addison, Capt. H. D. Pearson, R. E., Lieut. E. W. Cox, R. E., and the author. OCLC: 25320061.

Jones, A. E. Nyema and Stewart, William E. 1971. "General geology of Liberia." In: Conference on African geology, 1st. Pages 22-23. 1971. Commonw. Geol. Liaison Off., London. Conference on African Geology [abstracts of papers submitted on the occasion of the tenth anniversary of the Department of Geology, University of Ibadan, Nigeria, 7-14 December 1970. (1st : 1970: University of Ibadan) Organized by the Dept. of Geology. Descriptors: Africa; areal geology; economic geology; Liberia; petrology; sedimentary-petrology; West Africa. OCLC: 39782745.

Jolly, J. L. W. 1977. "The mineral industry of Liberia." Minerals Yearbook. 1974, Vol. 3; Area reports; international, Pages 607-615. 1977. U. S. Bureau of Mines. Washington, DC, United States. Descriptors: 1974; Africa; economic geology; economics; gas; Liberia; metals; mineral resources; natural; nonmetals; petroleum; production; West Africa; Economic geology-general. References: 12; tables. ISSN: 0076-8952.

"Jungle Iron." 1958. Mine and Quarry Engineering. February 1958. London. Volume 24, number 2, page 55. OCLC: 33909000.

Kaiser Engineers. 1962. Map of north-west Liberia. Kaiser Engineers International. Variant North-west Liberia. [Oakland, Calif.]: Kaiser Engineers International. Cartographic Material. Description: 1 map: photocopy; 54 x 38 cm. Scale 1:1,000,000. 1 cm. = 10 km. Notes: Covers Monrovia region. Relief shown by hachures. "J.V.T." "Job no. 6140-8." Subjects: Monrovia Region (Liberia)--Maps. LCCN: 96-687120.

Kartkonsult, AB. 1961. Liberia general map. Sanniquellie District. Stockholm: AB Kartkonsult maps : photocopies ; 50 x 51 cm. Scale 1:20,000. Language: English. "Copyright: AB Kartkonsult, Mapping Consultants Ltd., Stockholm." Notes: Relief shown by contours. "Aerial photography 1960." "Lamco Joint Venture Nimna Mining Project." "Liberian planimetric system." "Nimba geodetic network." Includes index map. OCLC: 50939388.

Kartkonsult, AB. 1964. Liberia general map. Sanniquellie District. Other Titles: Sanniquellie District. Stockholm: AB Kartkonsult. maps: photocopies; 50 x 51 cm. Scale 1:20,000. "Copyright: AB Kartkonsult, Mapping Consultants Ltd., Stockholm." Notes: Relief shown by contours. "Aerial photography 1960." "Lamco Joint Venture Nimna Mining Project." "Liberian planimetric system." "Nimba geodetic network." Includes index map. Subjects: Sanniquellie District (Liberia)- Maps, Topographic. LCCN: 2002-629013; LCCN: 2002-629014; OCLC: 50939388; OCLC: 50939399.

Kelly, Captain [sic]. 1867. "St. Paul's River, Liberia at its mouth." New York: Endicott & Co. Lith. Description: 2 maps on 1 sheet; each 24 x 37 cm., sheet 59 x 41 cm. Series: [American Colonization Society map collection; 18-19]. Subjects: Canals, Liberia. Planning Maps. Saint Paul River (Liberia), Maps. Notes: "The numbers indicate depth in fathoms, April 8th, 1867." Relief shown by hachures on canal map. Depths shown by soundings on St. Paul River map. Includes profile of "Trial section of canal along line no. 1."Available also through the Library of Congress Web site as a raster image. Map Info: Scale not given.; Scale [1:19,008]. 24 chains = 1 in. Surveyed by Captn. Kelly; drawn by H.R.W. Johnson. Survey of a route for a canal to connect the

Mesurado and Junk rivers in Liberia by Daniel B. Warner and H.R.W. Johnson, made in January & February 1867. LCCN: 96-684981; OCLC: 37986377; Access: http://hdl.loc.gov/loc.gmd/g8882s.lm000005

Kelly, Holzman, Michaelis & Associates. 1967. "Peace Corps/Liberian Medical Evacuation & Communication Network." Monrovia: Kelly, Holzman, Michaelis & Associates, cartographers. 1 map, photocopy, 53 x 49 cm. Scale not given. Notes: Shows hospitals, landing strips/airfields, roads, and distances between towns. "November 1967." Includes table of approximate flight times between landing strips, expressed in minutes. LCCN: 96-687116.

Kelley, L.; Spiker, E. and Rubin, M. 1978. "U. S. Geological Survey, Reston, Virginia, radiocarbon dates XIV." Radiocarbon. 20; 2, Pages 283-312. 1978. American Journal of Science. New Haven, CT, United States. Descriptors: absolute age; Africa; Atlantic Ocean Islands; Azores; C-14; carbon; Cenozoic; dates; Eastern U.S.; isotopes; Liberia; Mexico; Midwest; Pacific Ocean; Quaternary; radioactive isotopes; sediments; United States; West Africa; Western U.S. Quaternary geology. References: 69. ISSN: 0033-8222.

Khlestov, V. V. 1973. "Africa." In: The Facies of Regional Metamorphism at Moderate Pressures; A mineralogical-petrographical review of the principal regions of development of the bipyroxene-gneiss facies. Pages 82-87. 1973. Australian National University Press, Canberra. 299 p. illus. 26 cm. Translation of: Fatsii regionalnogo metamorfizma umerennykh davleniĭ. Added title page in Russian. Bibliography: p. 259-297. Descriptors: Facies (Geology); Rocks, Metamorphic; Africa; Algeria; Angola; Benin; bipyroxene gneiss facies; Central Africa; Central African Republic; Congo Democratic Republic; East Africa; facies; Ghana; granulite facies; Guinea; high temperature; Liberia; Malawi; Mali; Mauritania; metamorphic rocks; mineral assemblages; Mozambique; Nigeria; North Africa; petrology; Precambrian; shields; Sierra Leone; South Africa; Southern Africa; Tanzania; temperature; Togo; Uganda; West Africa; Zaire; Zimbabwe; Igneous and metamorphic petrology. ISBN: 0708107060; OCLC: 00897614.

Kirk, W. S. 1993. Minerals Yearbook, 1991: "Iron Ore. Annual report." Performer: Bureau of Mines, Washington, DC. April 1993. 38p. Notes: See also PB91-220541. Descriptors: commodities; International trade; exports; imports; production; Michigan; Minnesota; Missouri; Utah; Australia; Brazil; Canada; France; India; Iran; Liberia; USSR; Venezuela; tables, data; global aspects; *Iron ores; *Mineral economics. Abstract: Mr. Kirk became the commodity specialist for iron ore in 1992, and was previously responsible for cobalt, depleted uranium, hafnium, nickel, radium, thorium, and zirconium. The domestic survey data were compiled by Robin C. Kaiser, statistical assistant, Branch of Data Collection, and Henry F. Sattlethight, management analyst. The world production table was prepared by William L. Zajac, Chief, International Data Section. NTIS Number: PB93233377XSP.

Koala, F. K. 1983. "Apercu sur les formations Precambriennes de Haute-Volta et leurs mineralisations." Translated title: "The Precambrian formations of Upper Volta and their mineralization." In: Geology for development; mineral resources and exploration potential of Africa. Kogbe, Cornelius A. (editor). Journal of African Earth Sciences. 1; 3-4, Pages 363. 1983. Pergamon. London-New York, International. Conference: Sixth general conference on African geology; Geology for development; mineral resources and exploration potential of Africa. Nairobi,

Kenya. December 11-19, 1982. Language: French. Descriptors: Africa; Birrimian; Burkina-Faso; Eburnean; economic geology; Liberian; metal ores; mineralization; Paleoproterozoic; Precambrian; Proterozoic; stratigraphy; upper Precambrian; West Africa; Economic geology; geology of ore deposits; Stratigraphy. ISSN: 0731-7247.

Koch, Peter. 1971. Interpretation of air photographs of shifting cultivation, Liberia: An aid in agrographical analysis. In: Contributions to Land Use Survey Methods. Occasional Papers (World Land Use survey). Berkhamstead, Herts, UK: Geographic Publications. 19 pages. ISBN: 0900394013; OCLC: 292606

Koch, Peter. 1996. Brandrodungsfeldbau in Liberia (shifting cultivation, Wald, Feld, Wechselwirtschaft, SWA S.91). Geographica Helvetica. Jahr. 51. Nr. 1. 1996. p. 31-36. Maps. Descriptors: Agricultural methods- Liberia.

Kofron, Christopher P. 1992. "Status and habitats of the three African crocodiles in Liberia." Journal of Tropical Ecology. Vol. 8. Part 3. August 1992. Pages 265-273.

"Kongo." 1967. 1:50,000 topographic map: US Defense Mapping Agency. Map Sheet 2339 1, Series G744, Edition 1. AMRS, Inc. See: http://www.africaminerals.com/

Kopczynski, F. 1936. "Mapa Liberji." Warszawa: Druk Wojskowego Instytutu Geograficznego. Scale 1:2,000,000. In Polish. Notes: "Opracowat kpt. F. Kopczyński." Library of Congress, Geography & Map Division.

Kornhauser, Ben A. 1986. The mineral industry of Liberia. U. S. Bureau of Mines, United States. Pages: 5. 1986. Descriptors: Africa; economic geology; Liberia; mineral resources; West Africa. Notes: Preprint from Minerals yearbook, U. S. Bureau of Mines, for the years 1985, 1984 and 1983.

Kromah, Fodee. 1974. "The Geology and Occurrences of Iron Deposits in Liberia and the Impact of Mining on the Environment." Doctoral thesis, Cornell University. Ithaca, NY. 1974. Notes: Dissertations Abstracts International, Vol. 35, No. 2, p. 893B, 1974. Descriptors: Africa; areal geology; chemically precipitated rocks; economic geology; environmental geology; genesis; iron formations; iron ores; Liberia; metal ores; metamorphic-rocks; metasedimentary; mining; mining geology; open pit mining; ore deposits; pollution; Precambrian; processes; sedimentary rocks; surface mining; west; West Africa. Subjects: Iron mines and mining- Liberia. Iron ores- Liberia. Geology- Liberia. Notes: Vita. The 6 plates in pocket of original thesis are reproduced on leaves [182-193]. Includes bibliographical references (leaves 142-145). Reproduction: Photocopy of typescript. Ann Arbor, Mich.: University Microfilms International, 1977. 22 cm. OCLC: 3899162.

Kromah, Fodee. 1971. The mineralogy and petrography of the Nimba iron ore. Master's thesis, Michigan State University. East Lansing, MI, United States. Pages: 57. 1971. Includes bibliographical references (leaves 56-57). Two maps in pocket. Descriptors: Africa; economic geology; iron; Liberia; metals; mineralogy; Nimba; petrology; West Africa; economic geology; geology of ore deposits. OCLC: 24602321.

Kryatov, B. N.; Prokof-yev, S. S.; Makstenek, I. O.; Mamedov, V. I. and Khain, V. Ye. 1985. "Stages of tectonic development and metallogeny in the western Leone- Liberian Shield, western Guinea, and Guinea-Bissau." Geotectonics. 19; 6, Pages 460-472. 1985. American Geophysical Union. Washington, DC, United States. Descriptors: Africa; evolution; Guinea; Guinea Bissau; igneous activity; Liberia; Liberian Shield; metallogeny; structural geology; tectonics; West Africa. ISSN: 0016-8521.

Kühn, Stefan. 2000. "Beitrag zur Entwicklung einer geographischen Datenbasis der Landnutzung." Translated title: "Contribution for the development of a geographical database of land use." Universität Giessen, Department of Biology and Chemistry. Language: German. Notes: "Weltweit haben menschliche Aktivitäten den Zustand der terrestrischen Biosphäreverändert. Entwaldung und die Etablierung von Ackerland sind und waren die markantesten Manifestationen dieser Aktivitäten. In dieser Arbeit werden Datensätze erstellt, die digitalisierte Informationen über das Was, Wieviel, Wann und Wo von Landnutzungsänderungen enthalten. Die räumliche Auflösung der Daten beträgt 0.5 x 0.5 Grad und richtet sich nachden Erfordernissen dynamischer regionaler oder globaler Kreislaufmodelle, die einesolche Auflösung mehrheitlich verwenden.Publikationen, die auf der Basis von Satellitenaufnahmen oder historischen Landnutzungsdaten Landnutzungsänderungen geographisch und zeitlich rekonstruieren,bestenfalls in Form von Karten, sind systematisch ausgewertet worden, um die Datensätze zu erstellen. Für 12 Länder konnten Dateien über Landnutzungsänderungen erstellt werden:Brasilien, Paraguay, die Vereinigten Staaten, Costa Rica, Liberia, Ghana, Elfenbeinküste, Nigeria, Madagaskar, Malaysia, die Philippinen und Australien.Ein wichtiger Schwerpunkt lag auf der Quantifizierung und Lokalisierung der massiven Rodungen im Amazonasgebiet.Die vorgelegten Datensätze können in Verbindung mit dynamischen Modellen dabeihelfen, die Rolle von Landnutzungs- und Landbedeckungsänderungen in den biogeo-chemischen Kreisläufen besser einzuschätzen, insbesondere in quantitativer Hinsicht. Die Ergebnisdateien sind auf Anfrage vom Autor erhältlich." Institution: Bibliotheksservice-Zentrum Baden-Wurttemberg (BSZ BW), Germany, Virtueller Medienserver. Master's Thesis. Translated notes: World-wide human activities changed the condition of the terrestrial biosphere. The establishment fields from woodland are the most salient manifestations of these activities. In this work data records are provided, and it contains digitized information about how much, when and where are these changes in land use. The spatial dissolution of the data amounts to 0,5 x 0,5 degrees and depends on the requirements of dynamic regional or global cycle models. Publications, which took historical land use data changes of land use on the basis of satellites on or geographical and temporal reconstructions, at best in the form of maps, were systematically rated, in order to provide the data records. For 12 countries files of changes in land use could be provided: Brazil, Paraguay, the United States, Costa Rica, Liberia, Ghana, the Ivory Coast, Nigeria, Madagascar, Malaysia, the Philippines and Australia. An important emphasis was on quantification and localization of the substantial vegetation reduction in the Amazon area. The submitted data records can help in connection with dynamic models to estimate the role of land use and changes of land coverage in the biogeochemical cycles better in particular in quantitative regard. The resulting files are available on request of the author. Institution: Library service center Bad- Wurttemberg (BSZ German Federal Armed Forces), Germany, virtual server. URL: http://geb.uni-giessen.de/geb/volltexte/2000/314

Kukharenko, A. A.; Mikhaylov, B. M. and Orlova, M. T. 1971. "K mineralogii kimberlitov Liberiyskogo shchita (Zapadnaya Afrika)." Translated title: "Mineralogy of kimberlites of the

Liberian shield, west Africa." Sovetskaya Geologiya. 11; Pages 91-103. 1971. Notes: Vol. 11. Izdatel'stvo Nedra. Moscow, USSR. Language: Russian. Descriptors: Africa; chemical composition; distribution; igneous rocks; kimberlite; Liberia; Liberian Shield; mineral composition; petrology; plutonic rocks; trace elements; ultramafics; West Africa; Igneous and metamorphic petrology. Illustrations: sketch map. ISSN: 0038-5069.

Kulke, Holger. 1995. "Liberia, Sierra Leone." In: Regional petroleum geology of the world; Part II, Africa, America, Australia and Antarctica- Regionale Erdoel und Erdgasgeologie der Erde; Teil II, Afrika, Amerika, Australien und Antarktis. Beitraege zur Regionalen Geologie der Erde. 22, Part 2; Pages 127-128. 1995. Gebrueder Borntraeger Verlagsbuchhandlung. Berlin-Stuttgart, Federal Republic of Germany. 1995. Language: English. Descriptors: Africa; areal geology; Liberia; petroleum; petroleum-exploration; Sierra Leone; West Africa. ISSN: 0522-7038; ISBN: 3-443-11022-3.

Kümmerly and Frey. 1970. "Liberia." Bern, Switrzerland: Kümmerly and Frey. 1 color map, 50 x 49 cm.. Scale 1:1,100,000. Notes: Prepared for Mobil. Shows roads, railways, county and international boundaries, and Mobil Gasoline stations. Includes distance table. LCCN: gm71-1884.

Kümmerly and Frey. 1964. "Africa: Liberia- Roads." Produced for Mobil Oil. Scale 1:1,000,000. Library of Congress, Geography and Map Division.

Kunkel, G. 1966. "Anmerkungen über Sekundärbusch und Sekundärwald in Liberia (Westafrika)." Translated title: "Notes on Secondary Shrubs and Secondary Forest in Liberia (West Africa)." Plant Ecology. Volume 13, Number 4, July 1966. Pages: 233-248. Summary: The present abstract treats of the secondary formations of the tropical rain forest regions of Liberia. Modified high and swamp forests are analized with their particular succession sequences and compared with the so-called average-types of respective primary or well developed secondary formations. Gallery woods, savannah, mangrove and exploited forest are also discussed. The characteristics of secondary bush and woodland are easy to recognize. The results of investigations undertaken, which were of purely floristic character, confirm that: human influence (the shifting cultivation system) causes serious damage to the structure of natural forest formations; the composition of a specific area of forest is progressively impoverished if the regeneration-cultivation cycle is too short; typical secondary elements are not only more frequent than primary elements but are also characteristic in the floristic structure in general; if the underlying soil is liable to erosion continuous processes of forest degradation can lead to formation of savannah; repeated degradation of swamp forest often leads to an open type of herbaceous vegetation; certain "culture-companions" are conspicuous and serve as indicators for the identification and classification of the degradation or successive stages of such secondary formations. ISSN: 1385-0237.

Lacey, Linda. 1988. "Squatter settlement in Liberia: towards the integration of housing and population policies." African Urban Quarterly: Nairobi, Kenya. Vol. 3, Nos. 3/4. August and November 1988. Pages 219-229. Descriptors: Urban housing- Liberia.

Laenderbericht Liberia. 1989. Metzler-Poeschel. Stuttgart, Federal Republic of Germany. Pages: 85. 1989. Language: German. Descriptors: Africa; diamond; economics; gold ores; iron ores;

Liberia; metal ores; mineral-economics; native-elements; natural-resources; policy; production; West Africa. Illus. including 86 tables, sketch maps. ISSN: 0937-9967.

Lal, Rattan. 1993. "Soil erosion and conservation in West Africa." In: World soil erosion and conservation. Pimentel, David (editor). In the collection: Cambridge studies in applied ecology and resource management. 1993. Pages 7-25. 1993. Cambridge University Press. New York, NY, United States. 1993. Descriptors: Africa; Alfisols; Aridisols; Cameroon; compactness; conservation; degradation; economics; erosion; Inceptisols; Ivory Coast; land use; Liberia; Nigeria; nutrients; optimization; Oxisols; Sahel; savannas; Sierra Leone; soil erosion; soils; Ultisols; water erosion; water resources; West Africa; yields; Environmental geology; Soils. Illustrations, including 12 tables, sketch map. ISBN: 0-521-41967-0.

Lasserre, M. and Soba, D. 1976. "Age liberien des granodiorites et des gneiss a pyroxene du Cameroun meridional." Translated title: "Liberian age of granodiorites and pyroxene gneiss of southern Cameroon." Bulletin du Bureau de Recherches Geologiques et Minieres. Section 4: Geologie Generale. 1, Pages 17-32. 1976. Societe Geologique de France et la Bureau de Recherches Geologiques et Minieres. Paris, France. Language: French; Summary Language: English. Descriptors: absolute age; Africa; areal geology; Cameroon; chain silicates; dates; effects; geochronology; geologic maps; gneisses; granites; granodiorites; igneous rocks; isochrons; Liberian; maps; metamorphic rocks; metamorphism; plutonic rocks; polymetamorphism; Precambrian; Proterozoic; pyroxene group; Rb-Sr; silicates; south; upper Precambrian; West Africa; whole rock; Geochronology. Illustrations, including tables. ISSN: 0153-8446.

Latifi, Mohammad. 2004. Multinational Companies and Host Partnership in Rural Development: A Network Perspective on the Lamco Case. Doctoral thesis, Företagsekonomiska Institutionen, Uppsala Universitet, Uppsala, Sweden. Volume 113. (2006-05-01) Abstract: Multinational companies (MNCs) in less developed countries (LDCs) are regularly contracted to undertake rural development around their sites. Likewise, they regularly fail. How can a profit-making MNC encourage rural development in an undeveloped area? The purpose is to investigate how an MNC could fulfill its commitment in a way that benefits all involved parties. It is urgent and interesting to study how MNCs manage their relationships with non-business and business organizations in LDCs for local development. Starting points are community development and network theories. Non-business actors in business networks are focused, as this proved to furnish the most relevant description and analyze of the interaction of business and non-business organizations. The theoretical discussions explore the infusion of intermediary actors in order to bridge the gap between business and non-business actors. Beside some case studies for manifestation of this problematic issue, the study conducts two field studies at the site. This study follows ten years of Lamco Joint Venture Operating Company in Liberia which promised success in rural development, had it not been for drastically falling prices for iron ore and a civil war. Other MNCs have tried a one-directional way, whereas Lamco did co-operate in a network. The main result is that an intermediary PVO is an effective and efficient means for an MNC to fulfill its contractual commitments for rural development. The main theoretical contributions are the infusion of non-business intermediary actors to connect business and non-business actors, to enhance our understanding of relationships between MNCs and business and non-business actors in LDCs, and to understand the side-effects of business activities. The empirical contributions discuss the implications for MNCs, host governments, local communities and PVOs. See the site, 918Kb: http://urn.kb.se/resolve?urn=urn:nbn:se:uu:diva-4678 ISSN 1103-8454.

Lawrence, A. L. 1967. "Field prospecting for diamonds in isolated areas." Bulletin of the Geological, Mining and Metallurgical Society of Liberia. 2; Pages 1-3. 1967. Geological, Mining and Metallurgical Society. Monrovia, Liberia. Abstract: Acid gneiss and granitic intrusion bed rock, faults, Liberia. Descriptors: Africa; diamonds; economic geology; exploration; Liberia; west; West Africa; Economic geology, general. ISSN: 0367-4819.

Lawrence, A. L. 1965. "Field Prospecting in Isolated Conditions." Liberia Geological Survey. An unpublished un-numbered report. This appears to be a report by someone running a prospecting operation, may have been a talk at the Liberia Geological Society, 3 pages. AMRS, Inc. See: http://www.africaminerals.com/

Layzell, D. 1984. "A report on soil erosion in Liberia." FAO: Rome (Italy). 1984. 10 pages. Descriptors: Soil erosion, conservation and reclamation; Erosion; Soil Conservation; Liberia. Report No: FAO-AGO--GCP/RAF/181/NOR. Project: Regional Soil Conservation Project for Africa, Field document 20, RAF/181/NOR. Fiche No: 85W02455. Acc. No: 251921; MFN: 251921. FAO Library.

Lecorche, J. P. 1987. "Exotic terranes in the Mauritanide Orogen." In: International conference on Iberian terranes and their regional correlation. Spain, Universidad de Oviedo, Facultad de Geologia, Oviedo, Spain. Pages 86. IGCP (International Geological Correlation Programme). Conference: International conference on Iberian terranes and their regional correlation. Oviedo, Spain. September 1-6, 1987. Descriptors: Africa; Cenozoic; faults; IGCP; Liberia; Mauritania; Mesozoic; Morocco; North Africa; orogeny; Precambrian; Proterozoic; Sierra Leone; structural geology; tectonics; terranes; thrust faults; upper Precambrian; West Africa. Notes: IGCP Project No. 233.

Lecorche, J. P. 1982. "Structure of the Mauritanides." In: Regional trends in the geology of the Appalachian-Caledonian-Hercynian-Mauritanide Orogen; proceedings of the NATO Advanced Study Institute, Atlantic Canada. Schenk, Paul M., ed. Pages 347-353. 1982. D. Reidel Publ. Co.. Dordrecht, Netherlands. Conference: Regional trends in the geology of the Appalachian-Caledonian-Hercyanian-Mauritanide Orogen; NATO Advanced Study Institute. Fredericton, NB, Canada. August 2-17, 1982. Descriptors: Africa; fold belts; folds; Guinea; Hercynian Orogeny; Liberia; Mauritania; Mauritanides; Morocco; nappes; North Africa; orogeny; Paleozoic; Pan African Orogeny; Precambrian; Proterozoic; Senegal; Sierra Leone; structural geology; Taconic Orogeny; tectonics; upper Precambrian; West Africa; Western Sahara. References: 15; illustrations, including 1 table. ISBN: 90-277-1679-X.

LeFevre, Bernard Jr. 1966. A report on the geology and mining procedures at Bomi Hill, Liberia. Bulletin of the Geological, Mining and Metallurgical Society of Liberia. 1; 1, Pages 47-54. 1966. Geological, Mining and Metallurgical Society. Monrovia, Liberia. Abstract: Structural features, folding, regional metamorphism, lump ores, Precambrian. Descriptors: Africa; Bomi-Hill; economic geology; genesis; iron; Liberia; metals; mineral-deposits,-genesis; production; properties; structure; West Africa; Economic geology-of-ore deposits. ISSN: 0367-4819.

LeFèvre, Bernard. 1965. A report on the geology and mining procedures at Bomi Hills, Liberia. No place of publication given. Description: 7 leaves; 28 cm. Descriptors: iron ores- Liberia-Tubmanburg; iron mines and mining- Liberia- Tubmanburg. Notes: At head of Liberia Mining Company, Ltd. Cover title. Typescript. Bibliography: leaf 7. OCLC: 4061689.

Lemieux, H. J. (Director). 1970. A family of Liberia. Corp Encyclopaedia Britannica Educational Corporation. Visual Education Centre. [Toronto, Canada: Encyclopaedia Britannica Educational Corporation. Description: 1 videocassette (17 min.); sd., color; 1/2 in. Series: The African scene. Variation: African scene. Abstract: A documentary account of life in Liberia showing the contrast between the daily life of a mining-town family and the activities in a traditional Liberian village. Descriptors: Liberia- Social life and customs- Juvenile film; manuscript. System Info: VHS. Notes: Dubbed into video recording from the original motion picture. Responsibility: produced for Encyclopaedia Britannica Educational Corporation by Visual Education Centre; director, H. J. Lemieux. Camera, H. J. Lemieux. OCLC: 55525558.

Lemoine, Paul, 1878- and Parkinson, John, b. 1872. 1913. Afrique occidentale; English colonies on west coast of Africa and Liberia. Publication: Heidelberg, C. Winter. Description: 88 p. illustrations, (maps) folded map., diagms. 27 cm. Language: French. Series: Handbuch der regionalen geologie; heft 14. Descriptors: Geology- Africa, West; Geology- Liberia. Notes: "Bibliographie": p. 66-78. Par Paul Lemoine. English colonies on west coast of Africa and Liberia, by John Parkinson. OCLC: 1113297.

Lemoine, S. 1990. "Le faisceau d'accidents Greenville-Ferkessedougou-Bobodioulasso (Liberia, Cote d'Ivoire, Burkina Fasso), temoin d'une collision oblique eburneenne--The faults bundle Greenville-Ferkessedougou-Bobodioulasso (Liberia, Ivory Coast, Burkina Faso), evidence for an oblique Eburnean collision." In: 15th colloquium of African geology--15 (super e) colloque de geologie africaine. Publication Occasionnelle - Centre International Pour la Formation et les Echanges Geologiques = Occasional Publication - International Center for Training and Exchanges in the Geosciences. 20; Pages 29. 1990. Centre International pour la Formation et les Echanges Geologiques (CIFEG). Paris, France. 1990. Conference: 15th colloquium of African geology-15 (super e) colloque de geologie africaine. Nancy, France. September 10-13, 1990. Language: French; English. Descriptors: Africa; Burkina Faso; faults; Ivory Coast; Liberia; lineaments; northwestern-Ivory Coast; oblique orientation; Paleoproterozoic; Pan African Orogeny; plate collision; plate tectonics; Precambrian; Proterozoic; reverse faults; southern Liberia; tectonics; upper Precambrian; West Africa; western Burkina Faso; Structural geology. ISSN: 0769-0541

Lemoine, S. 1990. "Le faisceau d'accidents Greenville-Ferkessedougou-Bobodioulasso (Liberia, Cote d'Ivoire, Burkina Fasso), temoin d'une collision oblique eburneenne--The faults bundle Greenville-Ferkessedougou-Bobodioulasso (Liberia, Ivory Coast, Burkina Faso), evidence for an oblique Eburnean collision." In: Etudes recentes sur la geologie de l'Afrique; 15 (super e) colloque de geologie africaine; resumes detailles. Translated title: Recent studies on the geology of Africa; 15th colloquium on African geology; extended abstracts. Rocci, G (editor); Deschamps, M (editor). Publication Occasionnelle - Centre International Pour la Formation et les Echanges Geologiques = Occasional Publication - International Center for Training and Exchanges in the Geosciences. 22; Pages 67-70. 1990. Centre International pour la Formation et les Echanges Geologiques (CIFEG).

Paris, France. 1990. Conference: 15 (super e) colloque de geologie africaine. Nancy, France. Septembre 10-13, 1990. Language: French; Summary Language: English. Descriptors: Africa; Bobodioulasso Fault; Burkina Faso; faults; Ferkessedougou Fault; Greenville Fault; Ivory Coast; Liberia; lineaments; oblique orientation; Pan African Orogeny; plate collision; plate tectonics; Precambrian; Proterozoic; reverse faults; subduction; tectonics; upper Precambrian; West Africa; Structural-geology. Illustrations: References: 10; illustrations, including geol. sketch map. ISSN: 0769-0541.

Leo, G. W. 1967. "Age Investigations in Liberia." 15th Annual Report for 1967, Department of Geology and Geophysics, Massachusetts Institute of Technology. Cambridge, Mass. Pages 1-5.

Leo, G. W. 1967. "Geochronology program in Liberia." Bulletin of the Geological, Mining and Metallurgical Society of Liberia. 2; Pages 96. 1967. Geological, Mining and Metallurgical Society. Monrovia, Liberia. Descriptors: absolute-age; Africa; dates; granites; granitic; igneous rocks; Liberia; plutonic rocks; Rb-Sr; regional; West Africa; Geochronology. ISSN: 0367-4819.

Leo, G. W. and White, R. W. 1972. "Geologic Reconnaissance in Western Liberia." In: Continental Drift Emphasizing the History of the South Atlantic Area. Eos, Transactions, American Geophysical Union. 53; 2, Pages 175. 1972. American Geophysical Union. Washington, DC, United States. Descriptors: absolute age; Africa; areal geology; composition; dates; genesis; geochronology; igneous rocks; Liberia; metamorphic rocks; petrology; Precambrian; Rb-Sr; structural geology; structure; West Africa. ISSN: 0096-3941.

Leo, Gerhard W. and White, Richard W. 1967. Geologic reconnaissance in western Liberia. Washington, D.C.: U.S. Geological Survey. Description: 29 leaves: maps; 27 cm. Series: Open file report; Liberia investigations (IR); LI-13; Variation: Open-file report (Geological Survey (U.S.)); 67-147.; Interagency report.; Liberia investigations; 13. Subjects: Geology- Liberia. Notes: Open file numbering from publisher's list. "December 1967." One map on 1 folded leaf in pocket. Includes bibliographical references (leaves 25-29). US Geological Survey Inter-Agency Report, Liberia Investigations (IR LI 13), Referred to in press release dated December 29, 1967. OCLC: 52073399; USGS Library: (200) Un3ili no. 13.

Lersch, Johannes. 1970. "Montangeologische Grundlagen der betriebsgeologischen Abbausteuerung auf der Eisenerzlagerstaette Bong Range in Liberia Westafrika." Translated title: "Mining geologic principles involved in determination of the beneficiation properties of the iron ores in the Bong range, Liberia, west Africa." In: Montangeologische Untersuchungen an Lagerstaetten des Eisens, Mangans und Nickels. Clausthaler Hefte zur Lagerstaettenkunde und Geochemie der Mineralischen Rohstoffe. 9; Pages 108-140. 1970. Notes: No. 9. Borntraeger. Berlin-Stuttgart, Federal Republic of Germany. Language: German; Summary Language: English. Abstract: Low-grade ore (itabirite), geologic setting, exploration criteria, ore types, facies and genetic characteristics. Descriptors: Africa; Bong Range; economic geology; exploitation; iron; Liberia; metals; mineral deposits, genesis; occurrence; West Africa; Economic geology of ore deposits. Illustrations: including sketch map. ISSN: 0578-4697.

Lersch, Johannes. 1966. Geology of Bong Range iron ore deposit. Bulletin - Geological, Mining and Metallurgical Society of Liberia. 1; 1, Pages 13-21. 1966. Geological, Mining and

Metallurgical Society. Monrovia, Liberia. Abstract: Precambrian metamorphic rocks sequence itabirite, gneisses and lower mica schists, anticline and syncline structure, chemistry and mineralogy, west Leriul zones, Liberia. Descriptors: Africa; Bong-Range; economic geology; iron; Liberia; metals; properties; structure; West Africa; Economic geology-of-ore deposits. Illustrations: sketch maps. ISSN: 0367-4819.

Leuria, B. 1967. "Notes on the geology of the Lofa river south of Wea Sua, with a comment on diamond dispersal and recovery." Bulletin of the Geological, Mining and Metallurgical Society of Liberia. 2; Pages 4-8. 1967. Geological, Mining and Metallurgical Society. Monrovia, Liberia. Abstract: Geomorphology, predominance of acid gneiss series, three gravel types, terraces, Precambrian, Liberia. Descriptors: Africa; clastic sediments; diamonds; distribution; economic geology; gravel; gravels; Liberia; mining; placers; sediments; Wea Sua; West Africa; Yangaya; Economic geology of nonmetal deposits. ISSN: 0367-4819.

Leuria, B. 1966. "Diamond prospecting in Lofa County." Bulletin of the Geological, Mining and Metallurgical Society of Liberia. 1; 1, Pages 27-35. 1966. Geological, Mining and Metallurgical Society. Monrovia, Liberia. Descriptors: Africa; diamonds; economic geology; exploration; Liberia; Lofa County, Liberia; West Africa; Economic geology, general; sketch map. ISSN: 0367-4819.

Leuria, B. and Stracke, K. J. 1966. Diamonds and their occurrences in Liberia. Bulletin - Geological, Mining and Metallurgical Society of Liberia. 1; 1, Pages 5-12. 1966. Geological, Mining and Metallurgical Society. Monrovia, Liberia. Abstract: Distribution, source rocks, physical characteristics, mining. Descriptors: Africa; diamonds; economic geology; Liberia; production; properties; West Africa; Economic geology-general. ISSN: 0367-4819.

"Liberia." 1983. International Institute for Aerial Survey and Earth Sciences. Liberian Cartographic Service. [Monrovia]: Ministry of Lands, Mines and Energy, Republic of Liberia. Description: 1 map: color; 51 x 47 cm. Descriptors: Liberia- Maps, Topographic. Map Info: Scale 1:1,000,000; Transverse Mercator proj. International spheroid; (W 11030'--W 7020'/N 8045'--N 4010'). Coordinates--westernmost longitude: W0113000 Coordinates--easternmost longitude: W0072000. Notes: Alternate Liberia 1:1000000. Relief shown by contours and spot heights. Includes index diagram and index to adjoining sheets. Other Titles: Liberia 1:1000000. Cartography: International Institute for Aerial Survey and Earth Sciences-ITC, The Netherlands and the Liberian Cartographic Service. OCLC: 13198647.

"Liberia." 1982. International Institute for Aerial Survey and Earth Sciences. Liberian Cartographic Service. [Monrovia]: Ministry of Lands, Mines and Energy, Republic of Liberia. Description: 1 map: color; 51 x 47 cm. Descriptors: Liberia- Maps, Topographic. Map Info: Scale 1:1,000,000. Notes: Relief shown by contours and spot heights. "Relief data incomplete." Legend in English and French. In upper margin: Liberia 1:1000000. Ancillary map: Index diagram = Tableau synoptique, scale 1:10,000,000. Includes adjoining sheets index and notes. Other Titles: Liberia 1:1000000. Responsibility: cartography, International Institute for Aerial Survey and Earth Sciences-ITC, the Netherlands and the Liberian Cartographic Service, 1982. Document Type: Map. OCLC: 18190023.

"Liberia." 1982. In the collection: "Mapa geologico de Costa Rica." 1982. Costa Rica, Direccion de Geologia, Minas y Petroleo, San Jose, Costa Rica. 2CM-1, 1982. Minist. Ind., Energia y Minas. San Jose, Costa Rica. Pages: 1 sheet. 1982. Language: Spanish. Descriptors: Africa; areal geology; Central America; color geologic maps; Costa Rica; geology; Liberia; maps; West Africa. Map Coordinates: LAT: N103000; N112000; LONG: W0850000; W0861500. Map scale: 1:200,000.

"Liberia."1978. Annual Review of Mining. 1978; Pages 499. 1978. Mining Journal. London, United Kingdom. Descriptors: 1977; Africa; economic geology; economics; iron ores; Liberia; metal ores; mineral resources; ore deposits; production; review; West Africa; Economic geology of ore deposits. ISSN: 0026-5225; OCLC: 02438984.

"Liberia: L'Exploitation des minerais de fer du Liberia." 1952. La Chronique des Mines Colonials. Paris: Le Bureau d'études géologiques et minères coloniales. Language: French. 20e année, numbers 195-196. Page 221. OCLC: 7879812.

"Liberia: Mining Annual Review." 2000. Mining Journal. CD-ROM. ISSN: 0026-5225; OCLC: 02438984.

"Liberia." 1978. U. S. Geological Survey Professional Paper. Report Number: P1100. Pages 343-344. 1978. US Geological Survey. Reston, VA, United States. Descriptors: Africa; dikes; environmental geology; geologic maps; intrusions; Jurassic; Liberia; maps; Mesozoic; structure; West Africa; Environmental geology. ISSN: 1044-9612.

"Liberia." 1972. USGS Miscellaneous Investigations Series, No. I- . Notes: 1:250,000 in 10 sheets. A. Base (Shaded relief in color) B. Geology. C. Aeromagnetic. D. Radiometric. E. Gravity (only 1 quad) F. Minerals. Held in USGS Library Map Collection.

"Liberia." 1971. Encyclopaedia Britannica Educational Corporation. Encyclopaedia Britannica Educational Corp., 1971. Made by Visual Education Centre, Toronto. Description: 73 fr.; color.; 35 mm.; and phonodisc: 2 s., 12 in., 33 1/3 rpm., 12 min. Series: Africa: tradition and change. Abstract: Presents traditional village life in Liberia and compares it to life in the modern urban center at Monrovia. Explains that many of the inhabitants of Liberia are American blacks. Emphasizes the importance of foreign industry, and describes the rubber industry and the new iron mining industry. Descriptors: City and town life- Juvenile film; Industries- Liberia- Juvenile film; city and town life- Juvenile film; Liberia- Description and travel. Notes: Also issued with phonotape in cassette. With teacher's guide. Pre-adolescent; Filmstrip. LCCN: 72-734519; OCLC: 5002197.

"Liberia." 1969. No author given. Washington: No publisher given. November 1969. 1 color map, 13 x 13cm. Scale 1:4,000,000. Notes: "77251." LCCN: gm70-3349.

"Liberia." 1963. No author given. Monrovia? 1963? 1 map, photocopy, 34 x 34cm. Scale 1:1,500,000. Notes: Relief shown by hachures. LCCN: 96-687117.

"Liberia." 1962. Hollywood, CA: Films and Slides. Description: 1 filmstrip (26 fr.): color; 35 mm. Abstract: Depicts three styles of life (rubber plantation, city, and mining community) in the country of Liberia, the first republic on the continent of Africa. Descriptors: Liberia- Pictorial works. OCLC: 16320296

"Liberia." 1958. Mining Journal Annual Review. Page 163.

"Liberia." 1957. Government of Liberia. Map scale 1:1,000,000. 1 sheet. Notes: Plane coordinates on Hotines Rectified Skew Orthomorphic Projection. Geographic names by the board of Geographic Names of the Republic of Liberia. National boundaries not shown. No field check. Published in 1957 by the Government of Liberia through the Joint Liberia-United States Commission for Economic Development; the United States Coast and Geodetic Survey; the Liberian Cartographic Service and the Bureau of Mines and Geology, Treasury Department, Republic of Liberia. Compiled from shoran controlled mosaics of aerial photographs of 1953 by Aero Service Corporation, Philadelphia, Pa., USA. "497B." Library of Congress, Geography & Map Division.

Liberia. Bureau of Hydrocarbons. 1981. "Distribution of Petroleum in Liberia." Monrovia: The Bureau. 1 map, photocopy, 52 x 65 cm. Scale 1:920,000. Department of Minerals Exploration and Research, Ministry of Lands, Mines and Energy. LCCN: 85-690722.

Liberia. Bureau of Natural Resources and Surveys. Annual report on the activities of the Bureau of Natural Resources and Surveys. Other Titles: Annual report of the activities of the Bureau of Natural Resources & Surveys. Annual report for the period. Monrovia: Bureau of Natural Resources and Surveys. 28 cm. Notes: Title varies slightly. Description based on: 1969/1970. Report year ends Aug. 31. Descriptors: Mines and mineral resources- Liberia. ISSN: 1010-5476. LCCN: 75-640320 //r88.

Liberia. Department of Highways. Unknown date. "Map of Liberia." Notes: "Roads" Scale: 1:500,000. USGS Library- Map Collection.

Liberia. Department of Planning & Economic Affairs. 1970. "Transportation Map of Liberia." Monrovia: The Department. Color map, 56 x 43cm. Scale 1:1,100,000. Notes: "Includes distance chart." LCCN: gm72-2699; gm72-3700.

Liberia. Department of Public Works & Utilities. 1958. "Liberian Highway System: Road Mileage Checked and the Proposed System of National Primary and Secondary Roads." Monrovia: The Department. 1 map, photocopy, 52 x 56 cm., on sheet 77 x 87 cm. Scale 1:986,655. Notes: Also shows airfields and harbors. Drawn by E. R. Tutu. Recommended…, Chief of the Division of Highways, November 3, 1959. Includes distance chart. LCCN: 97-680004.

Liberia, Department of Public Works & Utilities. Office of National Physical Planning. 1968. "Map of Road Maintenance Districts." Monrovia: The Office. 1 map, photocopy, 37 x 37 cm. Scale 1:1,000,000. Notes: Also shows existing and proposed roads. Relief shown by hachures. LCCN: 96-68-468.

Liberia. Department of Public Works and Utilities. Office of Technical Planning. 1966. "Liberia: A Road Map." Monrovia: The Office. 1 map, photocopy, 39 x 39 cm. Scale 1:1,000,000. Notes: Shows existing and proposed roads. Relief shown by hachures. LCCN: 96-680473.

Liberia, Department of State. 1940. An act regulating the mining and prospecting of all minerals and other natural deposits within the Republic of Liberia; and Executive order no. 6, 1936, and amendment approved, February 20, 1940. Monrovia, Dept. of State. Description: 25 pages. Subjects: Mining law- Liberia. Power resources- Law and legislation. OCLC: 19718711.

Liberia. Division of Highways. 1957. "Liberian Roads and Population Densities." Monrovia: The division. 1 map, 20 x 40 cm. Scale 1:1,420,000. Notes: Also shows railroads, both existing and under construction. Mosaic data from Resources & Economic Development of Liberia. Joint US-Liberian Commission. 1957. Oriented with north toward upper left. LCCN: 96-687098.

Liberia. Ministry of Foreign Affairs. 1980. An act to amend an Act entitled "An Act to restore and control the diamond industry in Liberia amending and consolidating the laws regulating diamonds and for other purpose"- 1958 and to include the mining and prospecting of gold and other precious minerals or metals within the Republic of Liberia: approved September 21, 1979. Description: 11 pages; 24 cm. Descriptors: Diamond industry and trade – Law and legislation- Liberia. Notes: Cover title. "January 21, 1980." LCCN: 2001-427111; OCLC: 52160303.

Liberia. Ministry of Lands, Mines and Energy. 1994. "Petroleum Exploration Opportunities in the Republic of Liberia." Ministry of Lands, Mines & Energy unpublished un-numbered publication, 13 pages. AMRS, Inc. See: http://www.africaminerals.com/

Liberia. Ministry of Lands, Mines and Energy. Unknown date (1980's). "Mineral Resources and Concession Map of Liberia." 1:1,000,000 scale map by Liberia Ministry of Lands, Mines and Energy blue line copy with no number or authorship. 1:125,000 scale map by Liberia Geological Survey blue line copy with no number, date or authorship. AMRS, Inc. See: http://www.africaminerals.com/

"Liberia. Mines & Minerals- Iron." 1941. USGS. Prepared in cooperation with the republic of Liberia. Scale 1:1,000,000. Notes: Blueline print. W. part of Liberia, showing principal iron deposits, strike rock formations, etc. Library of Congress, Geography & Map Division.

Liberia. Ministry of Lands and Mines. Monrovia: Ministry of Lands and Mines. 1972- ."Report of the activities of the Ministry of Lands and Mines for the period ..." Other Titles: Annual report. Descriptors: Liberia. Ministry of Lands and Mines; Mines and mineral resources-Liberia; Periodicals; Surveying- Liberia. Periodicals. ISSN: 0304-7296; LCCN: 75-644089; OCLC: 2245983.

Liberia. Ministry of Lands, Mines, and Energy. 1980- . Annual report for the period ... [Monrovia, Liberia]: The Ministry. 28 cm. Notes: Description based on: Jan. 1, 1988 to Dec. 31, 1988; title from cover. Descriptors: Liberia. Ministry of Lands, Mines and Energy Periodicals; Mines and mineral resources Liberia. Periodicals. LCCN: 89-644120; OCLC: 10525887.

Liberia. Ministry of Public Works. Bureau of Operations. 1979. "Liberia Road Network." Monrovia: The Bureau. 1 map, photocopy, 119 x 95 cm. Scale 1:500,000. Notes: Prepared by SAUTI-ICE, consulting engineers, advisory services, road maintenance; assisted by Planning Division, M.P.W. Shows paved roads as of August 1979 and roads to be paved by June 1980. LCCN: 85-690726.

Liberia. Ministry of Public Works. Planning and Programming Bureau. 1982. "Highway Functional Classification System: Liberia." Monrovia: The Bureau. 1 map on two sheets, photocopy, 100 x 100 cm., sheets 125 x 61 cm. Scale 1:500,000. Notes: Traced by William C. Gray, approved by J. B. Freeman. February 26, 1982. LCCN: 85-690725.

Liberia 1:50,000. 1983. Liberian Cartographic Service. Great Britain; Directorate of Overseas Surveys. Great Britain. Ordnance Survey; Great Britain. Directorate of Overseas Surveys. Publication: Monrovia: Service [and] Directorate of Overseas Surveys, Southampton, England. Edition: Ed. 1-LCS/DOS. Description: maps: color; 74 x 59 cm. Or smaller on sheets 79 x 79 cm. or smaller. Series: Series LIB 50; Variation: LIB 50 (Series). Descriptors: Liberia- Maps, Topographic; Sierra Leone- Maps, Topographic. Map Info: Scale 1:50,000; Transverse Mercator proj. WGS 72 spheroid; (W 11030'--W 7015'/N 8045'--N 4015'). Notes: Relief shown by contours, spot heights and landform. Includes administrative boundaries map, index to adjoining sheets, magnetic declination diagram, grid zone designator and conversion table on each sheet. Includes parts of Guinea and Sierra Leone. Earlier sheets prepared jointly by the Government of Liberia (Liberian Cartographic Service) and the Government of the United Kingdom (Directorate of Overseas Surveys). Earlier sheets have the ed. statement: Ed. 1-LCS/DOS. "Published by the Government of the United Kingdom (Ordnance Survey) for the Government of Liberia (Liberian Cartographic Service)." Grid: UTM. Includes sheet history, conversion table, administrative boundary diagram, and index to adjoining sheets. Responsibility: prepared jointly by the Government of Liberia (Liberian Cartographic Service) and the Government of the United Kingdom (Directorate of Overseas Surveys). Other Titles: Some sheets titled: "Liberia 1:50,000 with part of Guinea"; Some sheets titled: "Liberia 1:50,000 with part of Côte d'Ivoire." Responsibility: prepared jointly by the Government of Liberia (Liberian Cartographic Service) and the Government of the United Kingdom (Ordnance Survey) under the UK Government's Technical Co-operation Programme. LCCN: 97-680175; OCLC: 16575384; 36857082.

Liberia: [mineral resources]. 1994. No author given. No place of publication given. 1 map, photocopy, 24 x 17 cm. Scale 1:3,000.000. Notes: Shows undeveloped mineral resources. Includes location map. LCCN: 97-687532.

Liberia: [mineral resources]. 1970. Liberian Cartographic Service. [Monrovia]: The Service. Description: 1 map: photocopy; 49 x 57 cm. Descriptors: Mines and mineral resources- Liberia- Maps. Map Info: Scale ca. 1:986,666; (W 11030'--W 7030'/N 8030'--N 4030'). Notes: "November 1956." Blue line print. Includes table of "Liberia grand total suplus [sic] commodities." Responsibility: Liberian Cartographic Service. LCCN: 96-687104; OCLC: 36191221.

"Liberia Builds First Railroad." 1950. Business Week. March 25, 1950, pages 136-139. Abstract: News note of a narrow gauge railroad line built by the Liberia Mining Company, and American enterprise by Republic Steel, operating under a new government concession at the Bomi iron mines. ISSN: 0007-7135.

"Liberia Gets Its First Power Dam." 1944. Business Week. 11 March 1944. Page 109. ISSN: 0007-7135.

"Liberia is Newest Foreign source for United States Iron." 1951. Business Week. 30 June 1951. Pages 24-25. ISSN: 0007-7135.

"Liberia: Road Classification Map." 1981? Monrovia: no publisher given. 1 map, photocopy, 50 x 47 cm. Notes: In right upper margin: "Planning and Development Atlas." Annotated in pencil and various colored inks to show road classification. LCCN: 85-690712.

Liberia, Wirtschaftsdaten und Wirtschaftsdokumentation (Economic parameters and economic documentation). 1989. Bundesstelle fuer Aussenhandelsinformation (Federal Agency for foreign trade information). Cologne, Federal Republic of Germany. Pages: 2. 1989. Language: German. Descriptors: Africa; economics; Liberia; mineral economics; mineral resources; policy; West Africa; Economic geology, general, economics. ISSN: 0175-114X.

Liberia. Western Mining Corporation. 1989. An Act Approving the Agreement for a Joint Undertaking between the Republic of Liberia and WMC (Liberia) Limited. [Liberia?]: s.n. Description: 1 v. (various foliations): map; 30 cm. Language: English. Descriptors: Mining leases-Liberia. Named Corp: Western Mining Corporation (Liberia). Notes: Includes the text of the agreement. Other Titles: Act Approving the Agreement for a Joint Undertaking between the Republic of Liberia and Western Mining Corporation (Liberia) Limited. LCCN: 94-149062.

Liberia. Unknown date. "Transportation Map of Liberia." Scale: 1:1,100,000. USGS Library- Map collection.

"Liberia Expands Hydropower." 1990. ENR. Volume 224 (June 7 1990) page 27 et seq. Descriptors: Hydroelectric power plants- Construction; Dams- Construction; Electricity supply-Liberia. ISSN: 0891-9526.

The Liberia Mining Company: twenty years of growth and development. 1967. Monrovia: Mass Communications. 16 pages: ill. 28 cm. Named Corporation: Liberia Mining Company. Notes: Cover title. Includes bibliographical references. OCLC: 3008207.

"Liberia's Mineral Resources." Engineering and Mining Journal v. 186 (October 1985) p. 26. Descriptor: Mines and mineral resources- Liberia. ISSN: 0095-8948.

"Liberia's National Biodiversity Strategy and Action Plan." 2004. Abstract: The rocks of northern Liberia generally form part of the West Africa Cretan, recognized by its stability and general absence of tectonic activity during the last 2.5 billion years. This old and stable base was subsequently penetrated by younger rocks and then covered by metasedimentary and metavolcanic rocks of at least two younger tectonic events. The rocks of Liberian Age extend into neighboring Sierra Leone, Guinea, and Ivory Coast and predominately are highly foliated granitic gneisses exhibiting a regional foliation and structural alignment in a northeasterly direction. Major faults along sections of the Lofa and the St. John River are parallel to regional lithological units and have significantly influenced present topography. Massive unfoliated to weakly foliated granitic rocks exist over large areas in the extreme north of the country. Within the Liberian Age Province are Metasedimentary rocks, such as quartzites, amphibolites, pelitic schists and banded ironstones technically called itabarite. Granitic gneisses and the metasedimentary rocks have been intruded by numerous northwest trending diabase dikes. These are parallel to the coast and represent intrusive activity associated with the onset of continental break-up in Jurassic time. Rocks of Eburnean Age are restricted to southeast Liberia where they extend into the Ivory Coast. Their structural trend is similar to those of the Liberian Age Province but is more biotite rich. A major tectonic feature within rocks of the Eburnean Age province is the Dube shear zone. It intersects the coastline about 40km west of Harper and strikes a NNE direction into the Ivory Coast. It is 2 to 3km wide and has been delineated on the basic of outcrops, topography and magnetic data. Rocks of the Pan-African Age are found along the coast from northwest of Greenville in the southeast to Sierra Leone. Unlike the northeastern regional trends of both the Liberian and Eburnean Age Provinces, structural trends within the Pan-African Province generally are northwesterly and parallel to the coastline. The rock types in this province range from basic igneous to peletic rock metamorphosed to the granulite and amphibolite grades. The Post Pre-Cambrian rocks in Liberia outcrop principally along the low-lying coastal area between Monrovia and Buchanan. Two onshore, sediment-filled basins also are located along this section of the coastline: the Roberts Basin filled with sediments of the Farmington River formation and Paynesville sandstone, and the Bassa Basin filled with material from the St. John River Formation. Rocks found in Liberia have been of economic importance and should continue to be in the future. Crystalline Rocks (igneous and metamorphic) are used locally in the construction industry as roadbed materials in building construction and as foundation stones in building construction. Post Pre-Cambrian rocks are used in the building industry where beach and river sands form the major constituents in the manufacture of concrete blocks. See: http://www.biodiv.org/doc/world/lr/lr-nbsap-01-p1-en.doc

"Liberia's New Boundary." 1908. Washington: American Geographic Society. Bulletin. Volume 40, pages 21-22.

Liberian American-Swedish Minerals Company. 1961. Liberia. Liberian American-Swedish Minerals Company (LAMCO). Joint venture. Maps, scale 1:125,000. Notes: Coverage of Liberia of the middle third of Liberia from the coast northeastward across the country.

Liberian American-Swedish Minerals Company. 1960. LAMCO joint venture agreement between the Liberian American-Swedish Minerals Company and Bethlehem Steel Corporation, dated as of April 28, 1960. Liberia? : Govt. of the Republic of Liberia. 1 volume (various pagings); 27 cm. Descriptors: Mines and mineral resources- Liberia; Joint ventures-Liberia; Community development- Liberia. Named Corporations: LAMCO. Bethlehem Steel Corporation. LCCN: 00-691243; OCLC: 47001947.

Liberian Cartographic Service. 1977. "Transportation Map of Liberia." Monrovia: The Service. May 1977. 1 map, photocopy; 101 x 101cm. Scale 1:500,000. Notes: "Based on a planimetric map of Liberia. Includes distance chart. Series Number LCS-21-77. LCCN: 85-690696.

Liberian Cartographic Service. 1976. "Map of National Forests and Related Data." Prepared by the Liberian Cartographic Services, Ministry of Lands and Mines. March 1976. Scale 1:1,000,000. 1 map, photocopy, 56 x 57cm. Notes: Shows forest concession areas. Forests and related data supplied by the Ministry of Agriculture. Series: LCS-18-76. LCCN: 85-690740.

Liberian Cartographic Service. 1976. "Transportation Map of Liberia." Monrovia: The Service. May 1977. 1 map, photocopy; 71 x 57cm. Scale 1:500,000. Notes: "Based on a planimetric map of Liberia. Includes distance chart. Series Number LCS-21-77. LCCN: 85-690694.

Liberian Cartographic Services. 1974. "Map of Hospitals and Clinics in Liberia." Monrovia: The Service. 1 map, photocopy, 71 x 57 cm. Scale 1:1,000,000. Notes: Prepared by the Liberian Cartographic Service, Ministry of Lands & Mines. Information obtained from the Ministry of National Health and Social Welfare, Republic of Liberia. Series no. LCS-4-74. LCCN: 85-690698.

Liberian Cartographic Service. 1957. "Communication map, Liberia." [Monrovia]: The Service. Description: 1 map; 17 x 24 cm. Descriptors: Transportation- Liberia- Maps. Scale [ca. 1:2,400,000]. Notes: Shows roads, railroads, airfields, lighthouses, and radio stations. "Feb. 1957." Oriented with north toward upper left. Relief shown by hachures and spot heights. LCCN: 96-687118; OCLC: 36191276.

Liberian Cartographic Service. 1956. "Liberia [Mineral Resources]." Monrovia: Liberian Cartographic Service. 1 map, photocopy, 49 x 57 cm. Scale 1:986,666. Notes: November 1956. Blue line print. Includes table of Liberia grand total suplus [sic] commodities." LCCN: 96-687104.

Liberian Cartographic Service. 1956. "Liberia [Road Map]." Monrovia: Liberian Cartographic Service. 1 map, photocopy, 56 x 66 cm. Notes: Shows road classification, proposed roads and railroads. Proposed road information DPWU, Division of Highway and other sources. Original by LCS March 1955. Blue line print. LCCN: 96-687115.

Liberian Cartographic Service. 1956. Republic of Liberia Geographic Names. No place of publication given. 1950-1959? Description: 29 leaves; 27 cm. Descriptors: Names, Geographical- Liberia. Notes: "Reprinted in Liberia, November 1956, by Liberian Cartographic Service." Includes index. LCCN: 78-112121; OCLC: 6086847.

Liberian Cartographic Service. 1955. Yearly rainfall and average temperature: [Liberia]. Monrovia. Liberian Cartographic Service. Description: 1 map; 17 x 22 cm. Descriptors: Rain and rainfall- Liberia- Maps; Atmospheric temperature- Liberia- Maps. Scale ca. 1:2,902,400; (W11030'--W 7030'/N 8030'--N 4030'). Notes: Shows "average temperature for the years 1954-1955." Base map: Republic of Liberia. Liberian Cartographic Service, 1952. LCCN: 96-687105; OCLC: 36191226.

Liberian Cartographic Service. 1954. Hydrographic survey, Freeport of Monrovia, Liberia. [Monrovia]: The Service. Description: 1 map: photocopy; 69 x 81 cm. Descriptors: Harbors - Liberia- Monrovia- Maps. Scale 1:5,000. Notes: "September 1954." Depths shown by contours and soundings. Blue line print. Other Titles: Freeport of Monrovia, Liberia. LCCN: 97-680002; OCLC: 36337373.

Liberian Cartographic Service. 1952. "A Proposed System of National Highways: Republic of Liberia." [Monrovia]: The Service. Description: 1 map: photocopy; 48 x 64 cm. Descriptors: Highway planning- Liberia- Maps; Roads- Liberia- Maps. Scale 1:970,000. Notes: At head of Republic of Liberia, Department of Public Works & Utilities, Technical Cooperation Administration, Bureau of Transportation. Blue line print. LCCN: 96-687109; OCLC: 36191240.

Liberian Cartographic Service. 195- . "Liberian, Number of Huts per Photo Mosaic." Monrovia?: The Service. 1 map, 21 x 24 cm. Scale not given. LCCB: 96-687102.

Liberian Cartographic Service and [Liberian] Department of Public Works & Utilities. 1959. Liberian highway system: road mileage checked and proposed system of national primary and secondary highways. [Monrovia]: The Department. Description: 1 map: photocopy; 52 x 56 cm., on sheet 77 x 87 cm. Descriptors: Roads- Liberia- Maps; Highway planning- Liberia- Maps. Scale ca. 1:986,655. Notes: Also shows airfields and harbors. "16-10-59." "Drawn by E. R. Tutu." "Recommended ... Chief, Division of Highways, November 3, 1959." "Base map by Liberian Cartographic Service." Includes distance chart. LCCN: 97-680004; OCLC: 36337384.

Liberian Cartographic Service and U.S. Coast and Geodetic Survey. 1952. Preliminary survey of Cape Palmas Harbor, Liberia. [Monrovia?]: The Service. Description: 1 map; on sheet 90 x 121 cm. Descriptors: Harper (Liberia)- Harbor- Maps. Scale ca. 1:2,400; (W 7043'/N 4025'). Notes: Relief shown by contours, hachures, and bath. soundings. Other Titles: Cape Palmas Harbor, Liberia. Hydrography by Liberian Cartographic Service; topographic compilation by U.S. Coast and Geodetic Survey. OCLC: 10781181.

Liberian Geological Survey. 1982. "Diamond Producing Areas, Gold and Associated Mineral Occurrences in Liberia with Their Status of Exploration." 1:1,000,000 scale map by Liberia Geological Survey blue line copy with no number or authorship. AMRS, Inc. See: http://www.africaminerals.com/

Liberian Geological Survey. 1985. "Mineral Resources Map of Liberia." 1:1,000,000 scale map by Liberia Geological Survey, blue line copy with no number or authorship. With WMC annotations from 1989-90. AMRS, Inc. See: http://www.africaminerals.com/

Liberian Geological Survey. 1985. "Mineral Resources Map of Liberia." 1:1,000,000 scale map by Liberia Geological Survey copy with no number or authorship. Primary gold areas of Liberia outlined. AMRS, Inc. See: http://www.africaminerals.com/

Liberian Geological Survey. 1981-. Annual report. Monrovia: Liberian Geological Survey. 28 cm. Notes: Description based on: October 30, 1981; title from cover. Report year ends Aug. 31. Descriptors: Liberian Geological Survey, Periodicals.; Geological research, Liberia, Periodicals; Geology Periodicals. LCCN: 76-641319 //r86; OCLC: 2436513.

Liberian Geological Survey. Bulletin. Other titles: Bulletin of the Liberian Geological Survey. Bulletin- Republic of Liberia, Bureau of Natural Resources and Surveys, Geological Survey.

Monrovia: The Survey, 1967- . Language: English. Illustrations, maps; 23 cm. ISSN: 0459-2204; LCCN: 73-432442; OCLC: 5356681.

Liberian Geological Survey. Memorandum report – Liberian Geological Survey. Monrovia. Description: v.; illus.; 28 cm. Language: English. Descriptors: Geology- Liberia- Surveys. Document Type: Serial. OCLC: 48117550.

Liberian Geological Survey. 1985. "Mineral Resources Map of Liberia." Monrovia: The Survey. 1 map, photocopy, 96 x 92 cm. Scale 1:500,000. Revised May 1985. LCCN: 91-682167.

Liberian Geological Survey. 1979. "Mineral Resources Map of Liberia." Monrovia: The Survey. 1 map, photocopy, 48 x 49 cm. Scale 1:1,000,000. Notes: December 1979. LCCN: 85-690703.

Liberian Geological Survey. 1969. "Republic of Liberia: Mineral Resources Map." No place of publication given. January 2, 1969. 1 map 32 x 30 cm. Scale 1:1,750,000. Notes: Sinkor? Prepared by Liberian Geological Survey, Bureau of Natural Resources & Surveys. LCCN: gm69-2021.

"Liberian Gold Confirmed for Mano River." 1999. Mining Journal. Number 332, page 256. ISSN: 0026-5225; OCLC: 02438984.

Liberian Information Service. 1963. "Liberia, Agricultural Resources Map." Produced and printed by the Liberian Information Service, in co-operation with US/AID. [Monrovia]: The Service. Related Names: United States. Agency for International Development. Cartographic Material. Description: 2 maps on 1 sheet; sheet 36 x 27 cm. Scale not given. Contents: [Rubber, oil palm products, and palm fiber]- [Coffee, cocoa, and kola nut]. Descriptors: Agricultural resources-Liberia-Maps; Crops- Liberia- Maps. LCCN: 96-687112.

Liberian Information Service. 1963. Liberia: Mineral Resources. Monrovia: The Service. 1 map, color, 31 x 31 cm. Scale not given. Notes: Produced by the Liberian Information Service in cooperation with the Bureau of Natural Resources and Surveys. LCCN: 96-687114.

Liberian Mining. 1970- . London, The Townsman. Description: No. 1 (1970)-; v.; illustrations. Descriptors: Mines and mineral resources- Liberia- Periodicals; Mining engineering- Periodicals. Notes: "Journal of mining development, geology and new equipment in the Republic of Liberia." Serial. OCLC: 9691927.

"Lofa River plantation proposed dock: Site plan." 1968. B.F. Goodrich Liberia, Inc. Corporate Name: Stanley Engineering Company. Monrovia, Liberia and Muscatine, Iowa, 1968. Cartographic Material. Description: map 27 x 41 cm. Scale Information: Scale 1:600. Notes: Photocopy. "Figure 1." Oriented with north toward the lower left. Descriptors: Lofa River, Liberia-Maps. LCCN: 69-000728; OCLC: 27212467.

Lok, Joris Janssen. 2004. "ISIS is deployed to Liberia on board the LPD Rotterdam (integrated staff information system) (landing platform dock)." Jane's International Defense Review. Volume 37. July 2004: page 25. Abstract: The Royal Netherlands Army (RNLA)'s command-and-control information system integrated staff information system (ISIS) is deployed in a maritime context during the deployment of the Royal Netherlands Navy (RNLN) landing platform dock (LPD) to Liberia. The ISIS system designed to enable the users to maintain and update information about ongoing land operations on a digital map is presented.

Lombard, Jean and Furon, Raymond. 1963. "Carte Gèologique de l'Afrique." Association of African Geological Surveys. Paris: ASGA-UNESCO. United Nations Educational, Scientific and Cultural Organization. Natural resources research. Number 3. Notes: Sheet 4 covers Liberia. LCCN: map 64-306; OCLC: 506488.

Luehrmann, W. H. 1969. "Seismic profile across the Atlantic." In: Unconventional methods in exploration for petroleum and natural gas. Heroy, W. B. (editor). Pages 127-137. Southern Methodist University. Dallas, TX, United States. Abstract: Film, cross section of deep ocean floor and sub-bottom sedimentary layering, Trinidad, B.W.I. and Monrovia, Liberia. Descriptors: Atlantic Ocean; bottom features; geophysical methods; geophysical surveys; marine geology; oceanography; seismic methods; surveys; West Indies- Liberia; oceanography. Notes: A symposium under the auspices of the Institute for the Study of Earth and Man, Southern Methodist University. Illustrations.

Lyday, Phillis A. 198? "The mineral industry of Liberia." U. S. Bureau of Mines, United States. Pages: 6. Descriptors: Africa; economic geology; economics; Liberia; mineral resources; production; review; West Africa; Economic geology,-general,-economics. Notes: Preprint from Minerals Yearbook, 1978-79, U. S. Bureau of Mines. USGS Library.

Lynch, William Francis (1801-1865). 1853. "Maryland in Liberia." United States. Navy Dept. Publication: [Washington, D.C.: U.S. Senate. Description: 1 map; 39 x 64 cm. Series: Senate ex. doc.; no. 1, 1st sess., 33d Cong.; [American Colonization Society map collection; number 14]; Variation: Senate executive document (United States. Congress. Senate); 33rd Congress, 1st session, no. 1. Descriptors: Maryland County (Liberia)- Maps. Notes: Relief shown by hachures. Depths shown by soundings. Inset: Cape Palmas. Available also through the Library of Congress Web site as a raster image. Map Info: Scale [ca. 1:245,000]; (W 7050'--W 6040'/N 5000'--N 4020'). LCCN: 96-684992; OCLC: 37986398; Access Path: http://hdl.loc.gov/loc.gmd/g8883m.lm000016

Lynch, William Francis (1801-1865). 1853. "Republic of Liberia." United States. Navy Dept. Washington, D.C.: U.S. House. Description: 1 map; 70 x 108 cm. Series: House executive. documents.; no. 1, 1st session, 33 Congress; [American Colonization Society map collection; 9]; Variation: Ex. doc. (United States. Congress. House); 33rd Congress, 1st session, no. 1. Subjects: Coasts- Liberia- Maps. Notes: Shows Liberian coastal area. Relief shown by hachures. Depths shown by soundings. Insets: Monrovia and Cape Musurado- Junk River and Marshall- Edina and Grand Bassa- Cestos- Sangwin River- Sinou. Available also through the Library of Congress Web site as a raster image. Map Info: Scale [ca. 1:485,000]; (W 120--W 80/N 70--N 50). LCCN: 96-684989; OCLC: 37986391; Access Path: http://hdl.loc.gov/loc.gmd/g8882c.lm000001

Maasha, Ntungwa. 1983. "Erosion of the Monrovia beaches." In: Geology for development; mineral resources and exploration potential of Africa. Kogbe, Cornelius A. (editor). Journal of African Earth Sciences. 1; 3-4, Pages 364. 1983. Pergamon. London-New York, International.

Conference: Sixth general conference on African geology; Geology for development; mineral resources and exploration potential of Africa. Nairobi, Kenya. December 11-19, 1982. Descriptors: Africa; beaches; engineering geology; erosion; geomorphology; Liberia; Monrovia; processes; shore-features; shorelines; Wamba Town; West Africa; West Point Beach; Yatono; Geomorphology. ISSN: 0731-7247.

McGrail, David W. 1977. "Mechanisms of sedimentation on the continental shelf of Liberia and Sierra Leone." Notes: Diss. Abstr. Int., Vol. 37, No. 10, p. 4943B, 1977. Doctoral thesis, University of Rhode Island. Kingston, RI, United States. Pages: 129. 1976. Descriptors: Africa; Atlantic Ocean; continental shelf; Liberia; marine transport; mechanism; North Atlantic; oceanography; sedimentation; Sierra Leone; transport; West Africa.

McGrail, David W. 1976. "Mechanisms of sedimentation on the continental shelf contiguous to Liberia and Sierra Leone." Eos, Transactions, American Geophysical Union. 57; 4, Pages 268-269. 1976. American Geophysical Union. Washington, DC, United States. Conference: American Geophysical Union; 1976 spring annual meeting. Washington, D.C. April 12-15, 1976. Descriptors: Africa; Atlantic Ocean; continental shelf; deposition; distribution; flocculation; Liberia; marine transport; oceanography; precipitation; processes; sedimentation; sediments; Sierra Leone; suspended materials; transport; West Africa; Oceanography. ISSN: 0096-3941.

McGrew, Harry J. 1983. "Oil and gas developments in Central and Southern Africa in 1983." AAPG Bulletin. 68; 10, Pages 1523-1599. 1984. Notes: World energy development, 1983. American Association of Petroleum Geologists. Tulsa, OK, United States. Descriptors: Africa; Angola; Benin; Cabinda-Angola; Cameroon; Central Africa; Central African-Republic; Chad; Congo; Congo-Democratic-Republic; discoveries; East Africa; economic geology; economics; energy-sources; Ethiopia; exploration; Gabon; geophysical-methods; geophysical surveys; Ghana; gravity methods; Indian Ocean Islands; Ivory Coast; Kenya; Liberia; Madagascar; Mauritania; Mozambique; Nigeria; production; South Africa; Southern-Africa; surveys; West Africa; Zaire; Economic geology, economics of energy sources. Illustrations, including 15 tables, sections, sketch maps. ISSN: 0149-1423.

McMaster, R. L.; Betzer, P. R.; Carder, K. L.; Miller, L. and Eggimann, D. W. 1977. "Suspended particle mineralogy and transport in water masses of the West African Shelf adjacent to Sierra Leone and Liberia." Deep-Sea Research (1977). 24; 7, Pages 651-665. 1977. Pergamon. Oxford and New York, International. Descriptors: Africa; Atlantic Ocean; composition; continental shelf; East Atlantic; Liberia; marine environment; marine transport; mineral composition; oceanography; sea water; sedimentation; sediments; Sierra Leone; suspended materials; transport; West Africa. References: 36; illustrations, including tables, sketch map. ISSN: 0146-6291.

Magee, Alfred W., III. 1985. Depositional environments of late Precambrian sediments from Liberia, Sierra Leone, and Senegal, West Africa. Master's thesis, Old Dominion University. Norfolk, VA, United States. Pages: 276. 1985. Descriptors: Africa; clastic rocks; environment; Gibi Mountain Formation; glacial environment; glaciomarine environment; Liberia; lithofacies; Mali-Group; marine environment; Precambrian; sedimentary petrology; sedimentary rocks; sedimentation; SEM data; Senegal; Sierra Leone; Tabe Formation; textures; tillite; West Africa.

Subjects: Sediments (Geology)– Africa, West. Geology, Stratigraphic- Precambrian- Liberia. Geology, Stratigraphic- Precambrian- Sierra Leone. Geology, Stratigraphic- Precambrian- Senegal. Notes: Includes bibliographical references. References: 63; 14 plates. OCLC: 13787550.

Magee, Alfred W. and Culver, S. J. 1986. "Recognition of late Precambrian glaciogenic sediments in Liberia." Geology (Boulder). 14; 11, Pages 920-922. 1986. Geological Society of America (GSA). Boulder, CO, United States. Descriptors: Africa; ancient ice ages; clastic rocks; diamictite; environment; Gibi Mountain; Gibi Mountain Formation; glacial environment; glacial geology; glaciomarine environment; ice rafting; laminite; Liberia; marine environment; paleoclimatology; Precambrian; Proterozoic; sedimentary rocks; sedimentation; SEM data; stratigraphy; tillite; upper Precambrian; West Africa; west central Liberia; Sedimentary petrology; Stratigraphy. Map Coordinates: LAT: N063000; N064000; LONG: W0070000; W0071000. 19 references; illustrations, including geological sketch map. ISSN: 0091-7613.

Magee, Alfred W. and Culver, Stephen J. 1985. "Depositional environments of the late Precambrian Gibi Mountain Formation, Liberia, West Africa." In: The Geological Society of America, Southeastern Section, 34th annual meeting. Abstracts with Programs - Geological Society of America. 17; 2, Pages 121. 1985. Geological Society of America (GSA). Boulder, CO, United States. Conference: The Geological Society of America, Southeastern Section, 34th annual meeting. Knoxville, TN, United States. March 20-22, 1985. Descriptors: Africa; ancient ice ages; clastic rocks; diamictite; environment; Gibi Mountain Formation; glacial environment; glacial geology; glaciation; Liberia; Monrovia; mudstone; Precambrian; Rokelide Basin; sedimentary rocks; sedimentation; siltstone; stratigraphy; tillite; West Africa. ISSN: 0016-7592.

Mahncke, Karl Joachim. 1973. Methodische Untersuchungen zur Kartierung von Brandrodungsflachen im Regenwaldgebiet von Liberia mit Hilfe von Luftbildern. Munchen: Geographisches Institut der Universitat Munchen. 57 pages: ill., maps (2 fold.in pocket; 30 cm. Language: German, English. Notes: Summary in English. Bibliography: p. 51-57. Descriptors: Agricultural mapping- Liberia; Aerial photography in agriculture; Photographic interpretation. ISSN: 3920397673; LCCN: 74-334529; OCLC: 1194500.

Main Republik Liberia: Zentrale Wasserversorgung, Greenville. Auftraggeber: Bundesamt fur Gewerbliche Wirtschaft. Corporate Name: Gesellschaft fur Klaranlagen und Wasserversorgung. Published/Created: Mannheim [Ger., 1969?] Related Names: Germany (West). Bundesamt fur Gewerbliche Wirtschaft. Description: map 98 x 70 cm. Type of Material: Cartographic Material. Scale Information: Scale 1:2,000. Notes: Photocopy. "Datum: 7.5.1969." Relief shown by contours. Legend in German and English. "Bearbeiter: Dr. Thiele." Descriptors: Water supply- Liberia, Greenville. Maps. LCCN: gm 71001933.

Maley, J. 1987. "Fragmentation de la foret dense humide Ouest-Africaine et extension d'une vegetation montagnarde a basse altitude au Quaternaire recent; implications paleoclimatiques et biogeographiques." Translated "Fragmentation of the western African rain forest and the extension of a low altitude mountainous vegetation during the late Quaternary; paleoclimatological and bio-geographical implications." In: Paleolacs et paleoclimats en Amerique latine et en Afrique (20 000 and B.P.-actuel)--Paleolakes and paleoclimates in South America and Africa (20 000 years B.P.-present) by Martin, T. Y. Geodynamique. 2; 2, Pages 127-131. 1987. Office de la Recherche

Scientifique et Technique Outre-Mer (ORSTOM). Bondy, France. Conference: Paleolacs et paleoclimats en Amerique latine et en Afrique (20 000 ans B.P. - actuel). Bondy, France. January 29-30, 1987. Language: French and English. Descriptors: Africa; biogeography; Cameroon; Cenozoic; Central Africa; changes; Cross-River; forests; Gabon; Ghana; indicators; Ivory Coast; Lake Bosumtwi; Liberia; microfossils; mountains; Nigeria; paleoclimatology; palynomorphs; Quaternary; rain forests; stratigraphy; tropical environment; upper Quaternary; vegetation; West Africa. References: 3; strat. color, sketch maps. ISSN: 0766-5105.

Mamedov, V. I.; Bronevoi, V. A. (Bronevoy, V. A.); Makstenek, I. O.; Ivanov, V. A. and Pokrovskii, V. V. (Pokrovskiy, V. V.). 1983. "Subsurface water regime; the main controlling factor of the mineralogic and geochemical zonality of the crust of weathering on the Liberian Shield." Lithology and Mineral Resources. 18; 1, Pages 1-8. 1983. Consultants Bureau. New York, NY. Descriptors: Africa; areal studies; atmospheric precipitation; bauxite; chemical composition; chemically precipitated rocks; clay mineralogy; clay minerals; Fouta Djallon Plateau; gibbsite; goethite; groundwater; hydrochemistry; hydrology; infiltration; kaolinite; karst; laterites; levels; Liberia; Liberian Shield; mineral composition; oxides; pH; phase equilibria; resilicification; saturated zone; seasonal variations; sedimentary petrology; sedimentary rocks; sheet silicates; silicates; soils; surface water; surveys; unsaturated zone; water table; weathering; weathering crust; West Africa; West Guinea; zoning. References: 9; illustrations, including 1 table, sketch maps. ISSN: 0024-4902.

Maness, Lindsey Vance. 1993. Guinea, Guinea-Bissau, Sierra Leone and Liberia (most), partial Ivory Coast, Senegal and Mali: transportation. Golden, CO: Lindsey Vance Maness, Jr., Consulting Geologist, c1994. 1 map; 87 x 119 cm. MAP: Scale 1: 1,000,000; Lambert conformal conic proj. (W 17-W 6/N 13-N 5). Notes: Shows highways. OCLC: 32113095.

Manlan, K. 1986. Training for Water in Africa, (La Formation dans le Domaine de l 'eau en Afrique). Aqua. No. 3, p 150-153, 1986. Abstract: A survey of progress made in training by members of the African Union of Water Suppliers (L'Union Africaine des Distributeurs d 'Eau, UADE) is summarized. Each country is reviewed in turn and recommendations are made that might be taken individually or in a group. Countries from which information was obtained include Benin, Cameroon, Central African Republic, Guinea, Ivory Coast, Gabon, Liberia, Mali, Mauritania, Morocco, Niger, Togo, Tunisia, Senegal, and Zaire. Attention is drawn to the critical need for training in maintenance. Training methods are mentioned, and the balance between local, national and international efforts and particular priorities are examined. Descriptors: African Union of Water Suppliers; Training; Maintenance; Benin; Cameroon; Central African Republic; Guinea; Ivory Coast; Gabon; Liberia; Mali; Mauritania; Morocco; Niger; Togo; Tunisia; Senegal; Zaire; Policy.

Mano River Mine. 1969. National Iron Ore Company. Compiled and Drawn by W. G. J. M. Ghering. April 1969; Current: May 1969. 1 Map. Report number: G-10. SCALE: 1:1000. USGS Library- Map Collection.

Mano River Resources, Inc. 2000. "Geological resource of 768,000 contained ounces of gold independently estimated at Mano's King George Larjor and Weaju properties, at an average grade of 6.9 grams per tonne, 2000." Vancouver, Ontario, Canada, Mano River Resources, Inc., press release, July 11, p. 1.

Mano River Resources, Inc. 2000. "Kimberlite discovered in Mano's KPO range license, western Liberia, 2000." Vancouver, Ontario, Canada, Mano River Resources, Inc. press release, December 4, p. 1.

Mano River Resources, Inc. 2000. "Drilling at Mano's King George Larjor project increases gold resource, 2000." Vancouver, Ontario, Canada, Mano River Resources, Inc. press release, December 12, p. 1.

"Mano River's Gondoja Assays." 2000. Mining Journal, v. 334, no. 8584, May 26, p. 413. ISSN: 0026-5225; OCLC: 02438984.

"Map of Africa from the latest authorities: possessions of the European powers." Philadelphia: T. Cowperthwaite. 1850. Description: 1 map: color; 38 x 30 cm. Descriptors: Africa- Maps; Liberia-Maps. Map Info: Scale 1:50,000,000; (W 300—E 600/N 300--S 300). Notes: Inset: Map of the Republic of Liberia from Gurley's Report made to the Senate of the United States, 1850 [Scale 1:4,000,000]. "Entered according to act of Congress, in the year 1850, by Thomas Cowperthwaite & Co., in the Clerk's office of the District court to the Eastern District of Pennsylvania." OCLC: 40343898.

"Map of Liberia." 1892. Outward Bound Publishing Company. London. Scale 1:1,440,000. Notes: Includes territories ceded to Liberia by special treaties. Library of Congress, Geography and Map Division.

"Map of Liberia Prepared in the War College Division, General Staff." 1916. US Department of State. Military Attaché, Liberian Embassy. Scale: 1:500,000. Notes: "Improved road" "Trail (Foot)" "Military Post" "Americo-Liberian Settlement" "Heights in feet" "Authorities- Map prepared in the Office of the American Military Attaché from French and English Official Sources, and Route Sketches by Travelers and Officers of the Liberian Frontier Forces and US Hydrographic Charts." Library of Congress.

"Map of western Liberia showing Grand Cape Mount, portion of Lofa and Montserrado Counties." 1975. Liberian Cartographic Service. [Monrovia]: The Service. 1 map: photocopy; 89 x 45 cm. Descriptors: Grand Cape Mount County (Liberia)- Maps; Lofa County (Liberia)- Maps; Montserrado County (Liberia)- Maps. Scale 1:250,000; (W 11030'--W 10030'/N 8000'--N 6000'). Notes: Includes location map. "Series no./LCS-11-75." Other Titles: Western Liberia showing Grand Cape Mount, portion of Lofa and Montserrado Counties. Responsibility: prepared by Liberian Cartographic Service, Ministry of Lands & Mines, July 1975. LCCN: 85-690719; OCLC: 12836348.

"Maps of Liberia, 1830-1870." 2001. Note: "This collection of Liberia maps includes twenty examples from the American Colonization Society (ACS), organized in 1817 to resettle free black Americans in West Africa. These maps show early settlements in Liberia, indigenous political subdivisions, and some of the building lots that were assigned to settlers. This on-line presentation also includes other nineteenth century maps of Liberia. Select maps to view by subject, creator (mapmaker/publisher), title of map, geographic location, or keyword. "Maps of Liberia" is a part of the Library of Congress's American Memory project. Subjects: American Memory; American Memory Project; Colonization; Liberia; Liberia history; Manuscript maps; Maps; National Digital Library; National Digital Library Program; Liberia- Maps. URL: http://memory.loc.gov/ammem/gmdhtml/libhtml/libhome.html

Martin, G. 1982. Geologie des Kuestengebietes von Nordwest-Afrika suedlich der Sahara. Neue Erkenntnisse aus der Erdoelexploration. Translated title: Geology of Northwest Africa; new data on oil prospection in southern Sahara. Giessener Geologische Schriften. 30; 1982. Pages: 150. Lenz. Giessen, Federal Republic of Germany. Language: German; Summary Language: English. Descriptors: Africa; basin range structure; basins; Benin; Cenozoic; continental drift; continental margin; correlation; economic geology; Ghana; Guinea-Basin; Guinea-Bissau; Ivory Coast and Ghana Basin; Liberia Basin; Mali; marginal basins; Mesozoic; oil and gas fields; plate tectonics; Quaternary; Sahara; sections; Senegal; Senegal Basin; shear zones; stratigraphy; tectonics; Tertiary; theses; Togo; volcanism; volcanology; West Africa. Notes: Univ. Giessen, Giessen, DDR; Doctoral Thesis; BGR-1983 A 532. References: 90; illustrations, including geol. sketch maps, strat. cols. ISSN: 0340-0654.

Mascle, J. 1977. "Le Golfe de Guinee (Atlantique Sud); un exemple d'evolution de marges Atlantiques en Cisaillement." Translated title: "The Gulf of Guinea, South Atlantic; an example of the shear zone evolution of Atlantic margins." Memoires de la Societe Geologique de France. 128, Pages 102. 1977. Societe Geologique de France. Paris, France. IGCP (International Geological Correlation Programme). Language: French; Summary Language: English. Descriptors: Africa; Atlantic Ocean; Benin; evolution; fracture zones; geophysical methods; geophysical surveys; Ghana; Gulf of Guinea; Ivory Coast; Liberia; Nigeria; North Atlantic; ocean basins; oceanography; profiles; rift zones; seismic methods; South Atlantic; surveys; tectonophysics; Togo; West Africa; Solid-earth-geophysics. Notes: IGCP Project No. 058. References: 235; illustrations, including geological sketch maps. ISSN: 0369-2027.

Massaquoi, Momolu A. 1966. Inaugural address of A. Momolu Massaquoi, President of the Geological, Mining & Metallurgical Society of Liberia. Bulletin - Geological, Mining and Metallurgical Society of Liberia. 1; 1, Pages 1-4. 1966. Geological, Mining and Metallurgical Society. Monrovia, Liberia. Descriptors: 1964; Africa; associations; economic geology; Geological,-Mining-and-Metallurgical-Society-of-Liberia; history; Liberia; mineral resources; Monrovia; West Africa; Economic geology-general. ISSN: 0367-4819.

Master, Sharad. 2000. Bibliography of the geology and mineral resources of Liberia and Sierra Leone, and the adjacent Archaean terrains of Guinea and Cote d'Ivoire, West Africa. Information Circular, University of the Witwatersrand, Economic Geology Research Unit. 342; 2000. University of the Witwatersrand, Economic Geology Research Unit. Johannesburg, South Africa. Pages: 67. 2000. Descriptors: Africa; Archean; areal geology; bibliography; Cenozoic; Guinea; iron ores; Ivory Coast; Liberia; Mesozoic; metal ores; mineral resources; mining; Paleoproterozoic; Paleozoic; Pan-African Orogeny; Precambrian; Proterozoic; Sierra Leone; upper Precambrian; West Africa. Subjects: Geology- Liberia- Bibliography. Mines and mineral resources- Liberia-Bibliography. Geology- Sierra Leone- Bibliography. Mines and mineral resources- Sierra Leone-

Bibliography. Geology- Guinea- Bibliography. Mines and mineral resources- Guinea-
Bibliography. Geology- Côte d'Ivoire- Bibliography. Mines and mineral resources- Côte d'Ivoire-
Bibliography. Notes: "March, 2000." OCLC: 44159956; ISBN: 1868382788; ISSN: 0375-8087.

Master, S. 2000. "Bibliography of the geology and mineral resources of Liberia, Sierra Leone and
the adjacent Archaean terrains of Guinea and Cote d'Ivoire, West Africa." Africa Geoscience
Review. 7; 2, Pages 115-185. 2000. Descriptors: Africa; bibliography; Guinea; Ivory Coast;
Liberia; Sierra Leone; West Africa.

Mattick, R.E. (comp.). 2005. Assessment of the petroleum, coal, and geothermal resources of the
economic community of West African states (ECOWAS) region. August 21, 2005. Notes:
Approximately 85 percent of the land area of the ECOWAS (Economic Community of West
African States) region is covered by basement rocks (igneous and highly metamorphosed rocks) or
relatively thin layers of Paleozoic, Upper Precambrian, and Continental Intercalaire sedimentary
rocks. These areas have little or no petroleum potential. The ECOWAS region can be divided into
13 sedimentary basins on the basis of analysis of the geologic framework of Africa. These 13
basins can be further grouped into 8 categories on the basis of similarities in stratigraphy, geologic
history, and probable hydrocarbon potential. The author has attempted to summarize the petroleum
potential within the geologic framework of the region. The coal discoveries can be summarized as
follows: the Carboniferous section in the Niger Basin; the Paleocene-Maestrichtian, Maestrichtian,
and Eocene sections in the Niger Delta and Benin; the Maestrichtian section in the Senegal Basin;
and the Pleistocene section in Sierra Leone. The only proved commercial deposits are the
Paleocene-Maestrichtian and Maestrichtian subbituminous coal beds of the Niger Delta. Some of
the lignite deposits of the Niger Delta and Senegal Basin, however, may be exploitable in the
future. Published literature contains limited data on heat-flow values in the ECOWAS region. It is
inferred, however, from the few values available and the regional geology that the development of
geothermal resources, in general, would be uneconomical. Exceptions may include a geopressured
zone in the Niger Delta and areas of recent tectonic activity in the Benue Trough and Cameroon.
Development of the latter areas under present economic conditions is not feasible. Subject:
Petroleum; Coal, Lignite, And Peat; Geothermal Energy; Africa; Coal Deposits; Geothermal
Resources; Petroleum Deposits; Stratigraphy; Resource Assessment; Chad; Gambia; Ivory Coast;
Liberia; Mali; Mauritania; Niger; Nigeria; Resource Potential; Senegal; Sierra Leone; Upper Volta;
Developing Countries; Geologic Deposits; Geology; Mineral Resources; Resources. URL:
http://www.osti.gov/servlets/purl/5585730-6FFPTL/native/

Mauche, Reneé Suzanne. 1985. Petrogenesis of diabase dikes in Liberia, West Africa, based on the
isotope compositions of of [sic] oxygen and strontium. Description: 153 leaves. Subjects: Geology-
Liberia. Diabase- Liberia. Dikes (Geology)- Liberia. Petrogenesis- Liberia. Notes: Includes
bibliographical references (leaves 151-153). Dissertation: Thesis (M.S.)--Ohio State University,
1985. OCLC: 13804371.

Mauche, R.; Faure, G.; Jones, L.M. and Hoefs, J. 1989. "Anomalous isotopic compositions of Sr,
Ar and O and the Mesozoic diabase dikes of Liberia, West Africa." Contributions to Mineralogy
and Petrology. 101; 1, Pages 12-18. 1989. Springer International. Heidelberg-New York,
International. 1989. Descriptors: absolute-age; absorption; Africa; alkaline-earth-metals; Ar-40;
argon; chemical-composition; dates; diabase; differentiation; dikes; fractional-crystallization;

geochemistry; geochronology; igneous-rocks; intrusions; isotopes; K-Ar; Liberia; magmas; Mesozoic; metals; noble-gases; O-18-O-16; oxygen; petrology; plutonic-rocks; pollution; ratios; Rb-Sr; Sr-87-Sr-86; stable-isotopes; strontium; tholeiitic-composition; West Africa. Illustrations: References: 27; illustrations, including 13 anal., 3 tables, geol. sketch map. ISSN: 0010-7999.

Maugham, Reginald Charles Fulke, 1866-1956. 1929. Africa as I have known it: Nyasaland; East Africa; Liberia and Saenaegal [sic]. London, J. Murray [1929]. Description: 372 pages front., plates, portraits, folded maps. 22 cm. Subjects: Africa, Central Description and travel. University of Wisconsin- American Geographical Society Collection. Call Number: DT351 .M4.

Maugham, Reginald Charles Fulke, 1866-1956. 1920? The republic of Liberia: being a general description of the Negro republic, with its history, commerce, agriculture, flora, fauna, and present methods of administration by R. C. F. Maugham; with map and 37 illustrations. New York: C. Scribner. Description: 299 pages, 14 leafs of plates: illustrations, folded map, portraits. 23 cm. Notes: "Liberian national anthem," pages 294-296. Includes index. Descriptors: Liberia; Liberia Politics and government; Liberia Economic conditions. University of Wisconsin- American Geographical Society Collection. Call Number: DT624 .M3 1920b.

Mcmaster, Robert L. and Lachance, Thomas P. 1969. "Northwestern African Continental Shelf Sediments." Rhode Island Univ., Kingston. Graduate School Of Oceanography. Studies Supported By Contract With Office Of Naval Research. 11 P, 5 Fig, 11 References. Abstract: An investigation of the gravel, sand, silt, and clay components and carbonate content of the bottom surface sediments on the northwestern African shelf and upper slope reveals that sand, composed of biogenic material, quartz, and glauconite, is the most common sediment type. This sand is primarily of biogenic origin between Ifni and Cape Blanc (21 deg n) and immediately south of Cape Verde (15 deg n) whereas the wide Guinea shelf is covered with low carbonate quartz sand. Glauconite is common on the outer shelf and upper slope. Modern mid-shelf slit facies occur immediately south of the strait of Gibraltar, on the south side of Cape Ghir (30 deg n), between the Gambia and Geba rivers, and along the narrow Sierra Leone-Liberian shelf. Generally, illite, kaolinite, and montmorillonite are present in these facies with kaolinite predominant in the tropical areas. Modern mid-shelf facies are found where the shelf is narrow and the sediment supply is ample. Apparently bottom turbulence and current action are not sufficient to keep the outer shelf free of fine sediment accumulation. Where the shelf is wide, modern fine sediments are being deposited in estuaries and on the inner shelf. Also, a calcareous sediment type is found where the shelf has been tectonically stable and the sediment supply inadequate. Finally, carbonate values, less than 50%, occur where a modern shelf facies is being deposited or where clastic sediments were distributed during pleistocene low sea levels. Descriptors: *sedimentation; *stratigraphy; *coastal plains; continental shelf; sands; silts; clays; erosion; provenance; marine geology; sediment transport; stratification; carbonates; *Africa; sedimentary facies. Marine Geology, Vol 7, No 1, Pp 57-67, Feb 1969.

Medvedev, V. Ya. and Sherif, M. 1981. "Opyt primeneniya matematicheskoy statistiki pri korrelyatsii nizhneproterozoyskikh otlozheniy Vostochno-Yevropeyskoy i zapadnoy chasti Afrikanskoy platform." Translated title: "Application of mathematical statistics to correlation of the lower Proterozoic of the East European Platform and western African Platform." Izvestiya Vysshikh Uchebnykh Zavedeniy. Geologiya i Razvedka. 1981; 6, Pages 27-32. 1981. Ministerstvo

Vysshego i Srednego Obrazovaniya. Moscow, USSR. Language: Russian. Descriptors: absolute age; Africa; Baltic Shield; chemically precipitated rocks; chert; Commonwealth of Independent States; correlation; correlation coefficient; dates; Europe; Fennoscandia; Finland; geochronology; iron formations; jaspilite; Liberia; Liberian Shield; metamorphism; Paleoproterozoic; petrology; Precambrian; Proterozoic; Russian Fennoscandia; Russian Platform; Russian Republic; Scandinavia; sedimentary rocks; statistical analysis; stratigraphic-boundary; stratigraphy; tectonics; Ukrainian Shield; upper Precambrian; USSR; West Africa; Western Europe. ISSN: 0016-7762.

Mehra, Avichal; Anantharaj, Valentine; Payne, Steve and Kantha, Lakshmi. 1996. Demonstration of a Real Time Capability to Produce Tidal Heights and Currents for Naval Operational Use: A Cast Study for the West Coast of Africa (Liberia). Mississippi State Univ Stennis Space Center Center For Air Sea Technology. Subject Categories: Physical And Dynamic Oceanography. Technical note, Report Date: 24 MAY 1996. 54 pages. Report number: CAST-TN-96-2. Supplementary Note: Original contains color plates: All DTIC reproductions will be in black and white. Descriptors: *computerized simulation; *real time; *demonstrations; *oceanographic data; *tidal currents; *case studies; *Naval Operations; *height; *Liberia; computer programs; coastal regions; lessons learned; ocean surface; models; finite element analysis; forecasting; storms; surges; littoral zones; ocean models; sea level; offshore; West Africa. Identifiers: CURRENTSS (Colorado University Rapidly Relocatable Nestable Tides And Storm Surge), CURRENTSS Computer Program. Abstract: This report documents an existing capability to produce operationally relevant products on sea level and currents from a tides/storm surge model for any coastal region around the world within 48 hours from the time of the request. The model is ready for transition to the Naval Oceanographic Office (NAVOCEANO) for potential contingency use anywhere around the world. A recent application to naval operations offshore Liberia illustrates this. Mississippi State University, in collaboration with the University of Colorado and NAVOCEANO, successfully deployed the CURRENTSS (Colorado University Rapidly Relocatable Nestable Tides and Storm Surge) model that predicts sea surface height, tidal currents and storm surge, and provided operational products on tidal sea level and currents in the littoral region off south-western coast of Africa. This report summarizes the results of this collaborative effort in an actual contingency use of the relocatable model, summarizes the lessons learned, and provides recommendations for further evaluation and transition of this modeling capability to operational use. Limitation Code: Approved for public release. Citation Creation Date: 25 July 1996. DTIC Number: ADA310368; ProxyURL/Handle: http://handle.dtic.mil/100.2/ADA310368

Meijers, G. J. 1981. Rainfall data book of Liberia: from inception till 1980. Liberia. Ministry of Lands and Mines. Monrovia, Liberia: Liberian Hydrological Service. Description: 133 leaves: maps; 28 cm. Descriptors: Rain and rainfall- Liberia; Watersheds- Africa; West. Liberia- Climate. Notes: "August 1981." At head of Department of Mineral Exploration and Research, Liberian Hydrological Service, Ministry of Lands & Mines. OCLC: 43167636.

Metelkina, M. P. 1976. "Associations of Precambrian diamond-bearing conglomerates." International Geology Review. 18; 10, Pages 1194-1200. 1976. Winston & Son. Silver Spring, MD, United States. Descriptors: Africa; African Platform; Arabian Peninsula; Arabian Shield; Asia; Brazil; Brazilian Shield; deposits; diamonds; distribution; economic geology; global; Guianan Shield; Hindustan Platform; Indian Peninsula; Indian Shield; Liberian; platforms; Precambrian; resources; Sierra Leone; South African; South America; South American; West Africa; Economic geology of nonmetal deposits. Illustrations: References: 25; table, geological sketch map. ISSN: 0020-6814.

Mikhaylov, Boris Mikhaylovich. 1969. "Geologiya i poleznyye iskopayemyye zapadnykh rayonov Liberiyskogo shchita." Translated Title: "Geology and mineral resources of the western part of the Liberian shield." Trudy - Vsesoyuznyy Ordena Lenina Nauchno-Issledovatel'skiy Geologicheskiy Institut im A.P. Karpinskogo, Novaya Seriya. 167; 1969. Notes: No. 167. Vsesoyuznyy Ordena Lenina Nauchno-Issledovatel'skiy Geologicheskiy Institut im A.P. Karpinskogo. St. Petersburg, USSR. Pages: 179. Language: Russian. Abstract: Stratigraphic and structural units, magmatism, rock types (with chemical analyses), kimberlite province, weathering crust, structure, fault tectonics, principal resources (bauxite, iron, diamonds, gold), other mineral resources, bibliography. Descriptors: Africa; areal geology; igneous rocks; Liberia; Liberian Shield; magmatism; mineral resources; Precambrian; processes; stratigraphy; structure; West Africa. Illustrations, including colored geol. map. ISSN: 0459-0856.

Mikhaylov, V. A. 2002. "Geologicheskiye osobennosti mestorozhdeniy zolota Zapadnoy Afriki." Translated "Geological characteristics of gold deposits in West Africa." Geologiya Rudnykh Mestorozhdeniy. 44; 6, Pages 500-509. 2002. MAIK Nauka/Interperiodika. Moscow, Russian Federation. 2002. Language: Russian. Descriptors: Africa; Ghana; gold ores; Ivory Coast; Liberia; Mauritania; metal ores; Paleozoic; placers; Sierra Leone; West Africa; Economic geology, geology of ore deposits. MAP COORDINATES: LAT: N020000; N250000; LONG: E0240000; W0173000. Illustrations: 20 References; illustrations, including geological sketch maps. ISSN: 0016-7770.

Milek, Andrew. 1948. Report on geological expedition to Liberia, 1948. Petroleum Advisers, Inc. New York: Petroleum Advisers, Inc. Description: [7], 56 leaves, [7] leaves of plates: ill.; 28 cm. Descriptors: Petroleum Geology- Liberia; Geology- Liberia. Notes: Typescript (carbon copy). Issued as a reconnaissance survey for Petroleum Advisers, Inc. OCLC: 20228952.

Miller, G. C. 1971. Commercial Fishery and Biology of the Freshwater Shrimp, Macrobrachium in the Lower St. Paul River, Liberia, 1952-53. National Marine Fisheries Service, Miami, Fla. Tropical Atlantic Biological Lab. Available From The National Technical Information Service As Com-71-00655. In Microfiche. National Oceanic 2nd Atmospheric Administration Special Scientific Report--Fisheries No. 626, February 1971. 13 pages, 8 Figures, 7 Tables, 15 References. Abstract: A small trap fishery was conducted for the large commercial freshwater shrimp macrobrachium vollenhovenii. A smaller species, m. Macrobrachion was culled from the trap catch for the fishermen's private use. The estuarine fishery was seasonal (May to January), during the normal low salinity periods. Cost of raw tail meats to the consumer was over $1.00 (u.s.) per pound. Fishermen derived more than $7,500 from the fishery. The life cycles of these species are reviewed. Monthly length distributions indicated that the fishery was supported by age group one, which was replaced at the end of the season by age group zero. Descriptors: commercial shellfish; fisheries; shrimp; fish harvest; commercial fishing; aquatic population; freshwater shrimp; Macrobrachium Spp.; shellfish populations; St. Paul River; Liberia.

Milliman, J. D. 1972. "Sediments of the East Atlantic Continental Margin A Preliminary Report." Woods Hole Oceanographic Institution, Mass. Technical Report Reference No. 72-2, January 1972.

18 P, 10 Fig, 4 Ref. Nfs Grant Gx-28193. Notes: "International Decade of Ocean Exploration, National Science Foundation"- Cover. "January 1972." "Technical report." Includes bibliographical references (leaf 7). Funding: Supported by the National Science Foundation. Abstract: The petrology, provenance and history of surficial sediments on the continental margin of Africa were studied using approximately 1000 samples obtained from collections. All sediment studies have been limited to depths less than 500 meters- that is the continental shelf and the upper slope. Three broad depositional areas can be recognized: Gibraltar to Cape Verde, Cape Verde to Liberia, and Liberia to the Niger. Sediments on the shelf and upper slope from Gibraltar to Cape Verde are rich in carbonate, primarily because of the small amount of fluvial sedimentation. The carbonate assemblages are temperate to subtropical. Absence of chemical weathering in this arid climate results in the retention of large amounts of feldspar. South of Cape Verde the sediments become increasingly terrigenous as fluvial sedimentation increases. The carbonate assemblages are subtropical to tropical. The tropical rivers in this area drain chemically weathered terrain. The result is a dominance of quartz rich sediments. Shelf sediments to the south and east of Liberia are dominated by fluvial muds. Many rivers in this area are short and drain coastal hills and mountains that are composed of crystalline rocks. As a result, the sediments tend to be more felspathic than normally would be expected in such a tropical area. Organic content in these sediments is high, probably the result of coastal upwelling as well as the deposition of river-borne plant material. Descriptors: *sedimentation; *provenance; *coastal plains; *continental shelf; *Africa; mineralogy; mud; sands; silts; carbonates; organic matter; weathering; deserts; tropical regions. OCLC: 36723589.

Mills, Dorothy, Lady, b. 1889. 1926? Through Liberia by Lady Dorothy Mills. New York: Frederick A. Stokes Co., [1926?]. Description: 240 pages: illustrations, front. (map) plates, portraits 22 cm. Notes: Includes index. Subjects: Liberia Description and travel. University of Wisconsin- Milwaukee Libraries -American Geographical Society Collection.

"Mine map showing advances for the months of July, August and September, 1974." 1974. Engineering Department, Liberia Mining Company, Ltd. 1 sheet. Map scale: 1:1200. USGS Library Map Collection.

Mineral Resources map of Grand Gedeh County [Liberia]. 1974. Liberia. Geological Survey. Ministry of Mines and Lands. No place of publication stated: Monrovia? Notes: 1 sheet, photocopy. 1:250,000. Held in USGS Library Map Collection.

Mineral Resources map of Maryland County [Liberia]. 1974. Liberia. Geological Survey. No place of publication stated. Monrovia? Notes: 1 sheet, photocopy. 1:125,000. Held in USGS Library Map Collection.

Molbak, K.; Hojlyng, N.; Jepsen, S.; Gaarslev, K. 1989. "Bacterial Contamination of Stored Water and Stored Food: a Potential Source of Diarrhoeal Disease in West Africa." Epidemiology and Infection EPINEU Vol. 102, No. 2, p 309-316, April 1989. 5 tab, 8 ref. Abstract: The food and water hygiene in two Liberian communities was studied in a house-to-house diarrhea survey. The level of contamination with enterobacteria of drinking water stored in the households was significantly higher than at the water sources. Food hygiene standards were low, particularly in the urban slum where storage of cooked food for long periods led to bacterial multiplication at high levels. Infant foods were particularly heavily contaminated. It is concluded that when water supply programs are planned, the presence of other risk factors for water-related diseases should be investigated. To ensure maximum health benefits, water projects should, as a rule, be accompanied by other interventions. The hazardous practices of storing large quantities of drinking water in open containers, for example, as well as storage of cooked food, must be discouraged as must the potential dangers of bottle-feeding and unhygienic practices during weaning. Breastfeeding and hygienic handling of food should be encouraged. Descriptors: developing countries; public health; human diseases; drinking water; water supply; potable water; hygiene; bacteria; pathogens; enterobacter; Africa; Liberia.

Monge, U. Alfonso. 1975. "Estudios geologicos basicos para la instalacion de una fabrica de vidrio en Liberia." Translated title: "Basic geologic studies for the installation of a glass factory in Liberia." Informes Tecnicos y Notas Geologicas. 13; 56, 1975. Language: Spanish. Descriptors: Africa; Bagaces Formation; Cenozoic; Central America; chemical composition; clastic sediments; Costa Rica; deposits; economic geology; economics; geologic maps; glasses; igneous rocks; Liberia; Liberia Formation; maps; mineral composition; possibilities; Quaternary; sand; sediments; utilization; volcanic rocks; West Africa. Map scale: 1:100,000; geologic maps. References: 24. ISSN: 0590-7012.

Monge, U. Alfonso. 1975. Estudios geologicos basicos para la instalacion de una fabrica de vidrio en Liberia. San Jose, Costa Rico, Ministerio de Economia, Industria y Comercio, Direccion de Geologia, Minas y Petroleo. Language: Spanish. 55 leaves: ill. (some color), maps (some color); 27 cm. Notes: Part of illustrative matter folded in pocket. Bibliography: leaves 54-55. OCLC: 3057013; USGS Library.

Morris, John Lewis and Massuquoi, Jaiah. 1930. Liberia. No place of publication given. Scale: 1:500,00. Library of Congress, Geography & Map Division.

Mortimer, Robert A. 1996. "Liberian crisis." The Journal of Modern African Studies. Cambridge. Vol. 34, No. 2. June 1996. Pages 293-306. Descriptors: Political geography- Liberia.

Mosto, Onuoha K. 1985. "Rheological properties of parts of the African lithosphere and their geodynamic significance." Journal of African Earth Sciences. 3; 4, Pages 437-442. 1985. Pergamon. London-New York, International. Descriptors: Africa; Atlas Mountains; Cape Range; crust; East Africa; East African Rift; Eastern Desert; Egypt; electromagnetic methods; geodynamics; geophysical methods; geophysical profiles; geophysical surveys; granites; gravity methods; heat flow; igneous rocks; Kanoma Pluton; Kenya; Liberia; lithosphere; mantle; Nigeria; North Africa; Paleozoic; plutonic rocks; Precambrian; rheology; Sahon Gida Pluton; Sierra Leone; surveys; syenites; tectonics; tectonophysics; teleseismic-signals; temperature; viscosity; West Africa; West African Shield; Western Desert; Zambia; Solid earth geophysics. References: 27; illustrations, including 2 tables. ISSN: 0731-7247.

Moszynski, Peter. 2003. "Liberia faces a humanitarian catastrophe, warn aid agencies. (News)." British Medical Journal 327 (7409). August 2, 2003: page 245. Abstract: Humanitarian agencies are warning that cholera and dysentery could be out of control as conditions deteriorate in Liberia's beleaguered capital, Monrovia. A rebel assault on the city that has killed hundreds of civilians has left hundreds of thousands more without access to food or clean water. With 200,000 displaced people crammed alongside the city's one million inhabitants, there is concern that an outbreak of cholera could run unchecked amid the squalor and carnage of the siege. Medecins Sans Frontieres had been treating 350 patients a week until its cholera clinics were overrun by rebel forces in the latest attack, when the town's only water treatment plant was also destroyed. Health workers fear that conditions will continue to deteriorate until a proposed US/Nigerian peacekeeping force arrives. "The epidemic of cholera now raging through Monrovia will only worsen if water and sanitation services are not provided immediately," says Sam Nagbe, who works in Monrovia for Oxfam. "People here are really suffering, but as long as the fighting continues we are unable to help them. If peacekeepers do not come, there will be a doomsday scenario."

Mukhtar, O. M. A. 1988. "Soil/land resource surveys in Liberia: review and appraisal." No place of publication given. 1988. 33 pages. Descriptors: soil surveys and mapping; land economics and policies; soil resources; land resources; surveys; soil types; soil classification; land suitability; evaluation; Liberia. UNDP Project. Field Document. Report No: FAO-AGO--LIR/87/010. Project: Land Resources Assessment for Land Use Planning, Liberia, Working paper 1, LIR/87/010. Fiche No: 88X01450. Acc. No: 277527. MFN: 277527. FAO Library.

Murray, G. E., & Pshorr, P. B. 1986. Preliminary Report on Further Reconnaissance Investigation for Gold within the Fasama - Masawo - Zolowo Area, Zorzor District, Lofa County: Liberia Geological Survey unpublished un-numbered report, 7 pages. AMRS, Inc. See: http://www.africaminerals.com/

Mykhaylov, V. A.; Shcherbak, D. N. 2001. "The formations and gold metallogeny of the early Proterozoic greenstone structures." Mineralogicheskiy Zhurnal. 23; 5-6, Pages 13-24. 2001. Language: English. Abstract: Precambrian greenstone belts are one of the oldest types of active structures of the earth crust. Their research has a big theoretical and practical significance, because they are important metallogenic structures and a lot of mineral deposits are concentrated in their areas: gold, iron, copper, uranium etc. The greenstone belts are formed not only in Archaean, but also in Paleoproterozoic. West Africa is one of the typical regions of the Early Proterozoic greenstone belts. They form narrow volcanic ridges, surrounded by sedimentary basins. The main metallogenic period in West Africa for gold and base metals is Eburnean. Mineralization is controlled on a regional scale by the large shear faults and greenstone belts, accreted to the Archaean nucleus of the Leonian-Liberian Shield (the Man Shield) of the period 2.1-1.9 Ga. There are analogous cycles in some other Precambrian cratons, including Precambrian of the Ukrainian Shield, and we can suggest global importance and practical significance of the Early Proterozoic tectono-metallogenic cycle with regard to gold and other mineralization. Descriptors: Africa; Aldan-Shield; Angola; Arctic-region; Asia; Baltic-Shield; Birrimian; Brazil; Brazilian Shield; Canadian Shield; Central Africa; Commonwealth-of-Independent-States; Congo-Democratic-Republic; cratons; East Africa; Eburnean; Europe; Ghana; gold ores; granites; Greenland; greenstone-belts; Guyana-Shield; igneous rocks; Indian Peninsula; Indian Shield; Kasai; Kola Peninsula; Liberia; Liberian Shield; metal ores; metallogeny; metamorphic-belts; mineral-deposits,-genesis; North-America; Paleoproterozoic; plutonic rocks; Precambrian; Proterozoic; Reguibat Ridge; Russian Federation; Russian Platform; shields; Siberian Platform; South America; Tanzania; tectonics; Ukraine; Ukrainian-Shield; upper Precambrian; West Africa; Economic geology, geology of ore deposits; Structural-geology. Illustrations: References: 67; illustrations, ISSN: 0204-3548.

Mykhaylov, V.A.; Shcherbak, D.N. 2000. "Comparative characterization of the Precambrian of the West African Craton and the Ukrainian Shield." Mineralogicheskiy Zhurnal. 22; 5-6, Pages 11-17. 2000. Natsional'naya Akademiya Nauk Ukrainy, Institut Geokhimii, Mineralogii i Rudoobrazovaniya. Kiev, Ukraine. 2000. Language: English: summaries in Ukrainian and Russian. Abstract: The major events in the Precambrian of the West African Craton are connected with the Archaean Preeburnean stage which includes the Leonian (3.2-2.9 Ga) and the Liberian (2.9-2.6 Ga) cycles, and the Early Proterozoic Eburnean stage (2.2-1.8 Ga). The Birimian deposit which formed greenstone belts plays a leading role in the structure and metallogeny of this region. The main metallogenic events are also connected with this stage. The Proterozoic Eburnean metallogenic cycle in the West Africa is supposed to have lasted over the period about 150 Ma from 2120+ or -41 (Perkoa deposit) to 2001+ or -17 Ma (Poura deposit). The fact that there are analogous cycles in other Precambrian cratons, including the Precambrian of the Ukrainian Shield attests to a global importance and practical significance of the Early Proterozoic tectono-metallogenic cycle in terms of gold mineralization. Descriptors: Africa; Birrimian; Commonwealth-of-Independent-States; Europe; gold ores; greenstone-belts; Leonian Cycle; Liberian Cycle; metal ores; metallogeny; metamorphic belts; mineral deposits,-genesis; mineralization; Paleoproterozoic; Pan-African-Orogeny; Perkoa Deposit; placers; Poura Deposit; Precambrian; Proterozoic; Russian Platform; sulfides; terrestrial comparison; Ukraine; Ukrainian Shield; upper Precambrian; veins; West Africa; West African Shield. References: 35; 2 tables. ISSN: 0204-3548

N-Diaye-I-Sory. 1997. "Geological and mining realities and possibilities in Guinea." In: Geological Society of America, 1997 annual meeting. Abstracts with Programs - Geological Society of America. 29; 6, Pages 124. 1997. Geological Society of America (GSA). Boulder, CO, United States. 1997. Geological Society of America, 1997 annual meeting. Salt Lake City, UT, United States. October 20-23, 1997. Descriptors: Africa; aluminum-ores; bauxite; Birrimian; chromite-ores; diamonds; gems; gold ores; Guinea; Liberian Shield; iron ores; metal ores; metallogenic provinces; mineral deposits,-genesis; mining; mining geology; nickel-ores; Paleoproterozoic; possibilities; Precambrian; Proterozoic; upper Precambrian; West Africa; Economic geology,-geology of nonmetal deposits. ISSN: 0016-7592.

Nair, A. M and Dorbor, J. K. 1997. "Some rare mineral occurrences in Liberia, West Africa." Zambian Journal of Applied Earth Sciences. 11; 1, Pages 73-77. 1997. Geological Society of Zambia. Lusaka, Zambia. 1997. Descriptors: Africa; Bolola; Bomi Hills; euxenite; fergusonite; Gbarma; Gbarma Hills; Gondoja, Liberia; gorceixite; Guinean Shield; Kakata; Kemata; Kumgbor; Liberia; mineral data; mineralization; minerals; niobates; niobotantalates; optical properties; oxides; picroilmenite; Sam Davis Creek; stolzite; tantalates; tungstates; West Africa; West African Shield; X-ray data; Mineralogy of non silicates. References: 7; geological sketch map. ISSN: 1010-5913.

Nair, A. M. and Dorbor, J. K. 1990. "Industrial minerals of Liberia." Industrial Minerals, March 1990:137.

Nair, A. M.; Duokenel, G. L. and Dorbor, J. K. 1988 "Prospects for Diamond Along Liberia-Guinea Border." Liberia Geological Survey unpublished un-numbered report, 16 pages. AMRS, Inc. See: http://www.africaminerals.com/

"National Forests of Liberia." 1970? Monrovia. Scale: 1:1,800,000. 1 map on sheet, 33 x 41 cm. Photocopy. LCCN: gm71-1694.

"National Forests: Liberia." 1968. Liberia: Bureau of Forest and Wildlife Conservation. Monrovia. No publisher given. 1 map, photocopy, scale: 1:1,000,000; 51 x 51cm. "Notes: in right lower margin: "DA-BFC. W. Copper." LCCN: 96-680496.

"National Forests: Liberia." 1962. [Monrovia?: s.n., 1962?]. Cartographic Material. Description: 1 map: photocopy; 41 x 38 cm. Scale 1:1,000,000. Notes: Relief shown by hachures. Base map: Planimetric map, Liberia. Descriptors: Forest reserves-Liberia- Maps. LCCN: gm71-001933; LCCN: 96-680470.

National Iron Ore Company of West Africa. 1967. Mano River Mine. Compiled and Drawn by W. G. J. M. Ghering. 1 map. Report number: G-5. SCALE: 1:1200. USGS Library- Map Collection.

Nelson, Harold D. 1984. Liberia, a country study. 3rd ed., 1984. Washington, D.C: Headquarters, Dept. of the Army. 340 pages: ill., maps. Series: DA pam; 550-38. Area handbook series. Notes: "Rev. ed. of: Area handbook for Liberia. 1972"—Title page verso. "Research completed September 1984." Includes index. Bibliography: pages 303-321. LCCN: 85-7393; OCLC: 11971799.

Nelson, Roy Andrew, 1896- . 1952. Airborne radioactivity survey in Liberia, West Africa. Series: TEM (Geological Survey (U.S.)); Washington: U.S. Geological Survey. 10 leaves; 27 cm. 361. Bibliography: leaf 10. USGS Library: (200) T67rm no.361.

Newhouse, Walter Harry, 1897-; Thayer, Thomas Prence, 1907- and Butler, Arthur P. 1908-. 1944. Report of the geological mission to Liberia, December, 1943- May, 1944, by W. H. Newhouse, T. P. Thayer, and A. P. Butler, Jr. Washington, US Dept. of the Interior, 1944? US Geological Survey, prepared in cooperation with the Republic of Liberia. Washington, D.C.: U.S. Dept. of the Interior, Geological Survey. Description: 139 leaves, 4 folded plates: ill.; 28 cm. 5 folded maps in pocket. Descriptors: Geology- Liberia; Iron ores- Liberia. Notes: Typescript (carbon copy). "Prepared in Cooperation with the Republic of Liberia." Other Titles: Geological Mission to Liberia. OCLC: 18481001.

Newhouse, Walter Harry, 1897; Thayer, Thomas Prence, 1907 and Butler, Arthur Pierce, 1908. 1945. Preliminary report on iron resources at Bomi Hills, Liberia. Other Titles: Iron ore reserves at Bomi Hills, Liberia. Washington: U.S. Geological Survey, 1945. 22 leaves; 28 cm. Series: Open-file report (Geological Survey (U.S.)); Number 681. Notes: Tables. Photocopy (positive). Descriptors: Iron ores-Liberia. USGS Library: (200) R29o no. 681

Neybergh, H.; Laduron, D.; Martin, H. and Verkaeren, J. 1980. "The vanadiferous magnetite deposits of the Oursi region, Upper-Volta." Economic Geology and the Bulletin of the Society of Economic Geologists. 75; 7, Pages 1042-1052. 1980. Economic Geology Publishing Company.

Lancaster, PA, United States. Descriptors: Africa; Birrimian Orogeny; Burkina Faso; diagenesis; economic geology; gabbros; igneous rocks; Liberian Orogeny; magnetite; metal ores; orogeny; Oursi; oxides; periodicity; plutonic rocks; Precambrian; Proterozoic; syngenesis; upper Precambrian; vanadium ores; veins; West Africa; West African Shield. References: 23; illustrations, including table, geological sketch map. ISSN: 0361-0128.

Niger Valley Exploring Party. 1861. Official report of the Niger Valley Exploring Party. By M. R. Delany, Chief Commissioner to Africa. New York: T. Hamilton, 1861. 75 pages; 23 cm. Descriptors: Yoruba (African people); African Americans Colonization Africa; Liberia-Description and travel. Other: Delany, Martin Robison (1812-1885). University of Wisconsin, American Geographical Society Library - Rare Book Collection. Call Number: DT625 N54x 1861.

Nyema, Jones A. E. and Stewart, W. E. 1972. "General geology of Liberia." In: African Geology; Regional Geology. Pages 495-504. 1972. Univ. Ibadan, Dep. Geol., Ibadan, Nigeria. Descriptors: absolute age; Africa; amphibolites; areal geology; arkose; basement; chemically precipitated rocks; clastic rocks; complexes; conglomerate; Cretaceous; diabase; dikes; distribution; exploration; geochronology; gneisses; igneous rocks; intrusions; iron formations; iron ores; itabirite; Liberia; Mesozoic; metal ores; metamorphic rocks; mineral resources; Monrovia; occurrence; ore deposits; petrology; plutonic rocks; Precambrian; schists; sedimentary rocks; siltstone; southwest; stratigraphy; structural geology; tectonics; textures; West Africa. Notes: With discussion. Illustrations, including sketch map.

Ocean Industry. 1967. "Geotech runs 7,000-mile Atlantic seismic survey." Ocean Industry. Volume 2; 10, Pages 46-49. 1967. Gulf Publishing Company. Houston, TX, United States. Descriptors: Africa; Antilles; Atlantic Ocean; Caribbean-region; geophysical methods; geophysical surveys; Lesser Antilles; Liberia; seismic methods; South Africa; Southern Africa; surveys; Trinidad; Trinidad and Tobago; West Africa; West Indies; applied geophysics. ISSN: 0029-8026.

Odokiy, B. N. 1971. "Kory vyvetrivaniya severo-zapadnoy chasti Liberiyskogo shchita." Translated title: "Weathered crusts in the northwestern part of the Liberian Shield." In: Kontinental'nyye pereryvy i kory vyvetrivaniya Sibiri. Trudy Sibirskogo Nauchno-Issledovatel'skogo Instituta Geologii, Geofiziki i Mineral'nogo Syr'ya. 126; Pages 158-163. 1971. Notes: No. 126. Zapadno-Sibirskoye Kniznoye Izdatel'stvo. Novosibirsk, USSR. Language: Russian. Descriptors: Africa; bauxite; economic geology; environment; Liberia; Liberian Shield; occurrence; products; weathering; West Africa; economic geology, general. ISSN: 0583-1822.

Offerberg, Jan and Tremaine, John Winthrop. 1971. "Report on LAMCO Joint Venture's geological investigations in Liberia between Nimba and Lower Buchanan along the Railroad Concession area." Liberian Geological Survey; LAMCO Joint Venture. Nimba, Liberia: Liberian Geological Survey, 1961, reprinted 1971. 85 pages: tables; 27 cm. Notes: Reprint of the 1961 ed., published by W. Reklam and Caslon Press Boktr., Stockholm, Sweden. Bibliography: p. 14-15. OCLC: 6167045.

Okigbo, B. N. 1977. "Farming systems and soil erosion in West Africa." In: Soil conservation and management in the humid tropics. Greenland, D. J. and Lal, R. (editors). Pages 151-163. 1977. John Wiley & Sons. New York, N.Y. Conference: Soil conservation and management in the humid tropics. Ibadan, Nigeria. June 1975. Descriptors: Africa; agriculture; Benin; Cameroon; Central Africa; Congo Democratic Republic; conservation; erosion; erosion control; Gabon; Ghana; humid-environment; Ivory Coast; Liberia; Nigeria; Sierra Leone; soil management; soil surveys; soils; surveys; tropical regions; water erosion; West Africa. Illustrations: References: 36; tables. ISBN: 0471994731.

Onstott, T. C. 1982. "(super 40) Ar (super 39) dating and correlation of paleomagnetic poles from the Guayana and West African shields." In: AGU 1982 spring meeting; abstracts. Eos, Transactions, American Geophysical Union. 63; 18, Pages 453. 1982. American Geophysical Union. Washington, DC. Conference: AGU 1982 spring meeting. Philadelphia, PA, United States. May 31-June 4, 1982. Descriptors: absolute age; Africa; Ar-Ar; Bolivar, Venezuela; correlation; dates; Eburnean Province; geochronology; Imataca Complex; Liberia; magnetization; minerals; Northern Hemisphere; paleomagnetism; pole positions; Precambrian; remanent magnetization; South America; Venezuela; West Africa. ISSN: 0096-3941.

Onstott, T. C. 1981. Ar-40/Ar-39 dating and paleomagnetism of age-provinces of Liberia. Liberian Geological Survey, Sci. Assoc. Liberia, Liberia. Pages: 4. 1981. IGCP (International Geological Correlation Programme). Descriptors: Africa; Ar40- Ar39; Ar-Ar; argon; Bolivar, Venezuela; chemically precipitated rocks; Cuchivero Group; Falawatra Group; gneisses; granites; granulites; greenstone; Guayana Shield; igneous rocks; Imataca Complex; iron formations; isotopes; Kanaku Complex; Liberia; Mesozoic; metamorphic rocks; noble gases; paleogeography; paleomagnetism; Paleozoic; Parguaza rapakivi; Permian; petrology; plutonic rocks; radioactive isotopes; Roraima diabase; Roraima sandstone; schists; sedimentary rocks; South America; stable isotopes; stratigraphy; Supama Pastora Complex; Triassic; Venezuela; West Africa. Notes: IGCP Project No. 108/144.

Onstott, T. C. and Dorbor, J. 198? "Argon 40 39 and paleomagnetic correlations between the Liberian West African Shield and the Venezuelan Guayana Shield." Liberia Geol. Survey, Liberia. Pages: 3. 198?. IGCP (International Geological Correlation Programme). Descriptors: absolute age; Africa; amphibolites; Ar-40-Ar-39; Ar-Ar; Archean; argon; dates; geochronology; granites; Guyana Shield; igneous rocks; isotopes; Liberia; metamorphic rocks; noble gases; paleomagnetism; plutonic rocks; Precambrian; quartz-monzonite; radioactive isotopes; South America; stable isotopes; Venezuela; West Africa; West African Shield; Geochronology. Notes: IGCP Project No. 108/144. Illustrations: References: 5; illustrations, including sketch map.

Onstott, T. C. and Dorbor, J. 1987. "(super 40úAr (super 39úAr and paleomagnetic results from Liberia and the Precambrian APW data base for the West African Shield." Journal of African Earth Sciences. 6; 4, Pages 537-552. 1987. Pergamon. London- New York, International. Descriptors: absolute age; Africa; alternating field demagnetization; amphibolites; Ar-Ar; Archean; biotite; dates; demagnetization; feldspar group; framework silicates; Kalahari Shield; Liberia; magnetization; metamorphic rocks; mica group; Nimba County; paleomagnetism; Precambrian; Rb-Sr; schists; sheet silicates; silicates; tectonophysics; thermal demagnetization; West Africa; West African-Shield. References: 50; illustrations, including 3 tables, geol. sketch maps. ISSN: 0731-7247.

Onstott, T. C.; Hall, C. M. and York, D. 1983. "40Ar and 39Ar dating of paleomagnetic poles from a West African Archean metamorphic terrane." In: American Geophysical Union; 1983 spring meeting. Eos, Transactions, American Geophysical Union. 64; 18, Pages 215. 1983. American Geophysical Union. Washington, DC, United States. Conference: American Geophysical Union; 1983 spring meeting. Baltimore, MD, United States. May 30-June 3, 1983. Descriptors: absolute age; Africa; amphibole group; amphibolites; Ar-Ar; Archean; biotite; chain silicates; clinoamphibole; dates; evolution; feldspar group; framework silicates; geochronology; granulites; hornblende; Ivory Coast; Liberia; lithostratigraphy; metamorphic rocks; mica group; Nimba County; orogeny; paleomagnetism; Pan African Orogeny; pole positions; Precambrian; Proterozoic; Rb-Sr; sheet silicates; silicates; thermal history; upper Precambrian; West Africa; Geochronology; Isotope geochemistry. ISSN: 0096-3941.

Onstott, T. C.; Pringle, Goodell L.; Henne, R.; King, B. Y. and Olds, P. 1986. "Argon degassing of hornblende, biotite and muscovite; hydrothermal vs. in vacuo." In: AGU 1986 fall meeting and ASLO winter meeting. Eos, Transactions, American Geophysical Union. Volume 67; 44, Pages 1249. 1986. American Geophysical Union. Washington, DC, United States. Conference: AGU 1986 fall meeting and ASLO winter meeting. San Francisco, CA, United States. December 8-12, 1986. Descriptors: Adirondack Mountains; Africa; amphibole group; Ar40; Ar39; argon; Benson Mine; biotite; chain silicates; clinoamphibole; degassing; Harper; high temperature; hornblende; isotopes; Liberia; low temperature; mica group; muscovite; New York; noble gases; radioactive isotopes; sheet silicates; silicates; stable isotopes; temperature; United States; West Africa; Geochronology. ISSN: 0096-3941.

Onstott, T. C.; Hargraves, R. B.; York, Derek and Hall, Chris. 1984. "Constraints on the motions of South American and African shields during the Proterozoic; I, (super 40) Ar (super 39) Ar and paleomagnetic correlations between Venezuela and Liberia." Geological Society of America Bulletin. 95; 9, Pages 1045-1054. 1984. Geological Society of America (GSA). Boulder, CO, United States. Abstract: The Encrucijada Pluton of Venezuela and amphibolites near Harper, Liberia, are located on opposite sides of the Liberian Pan-African mobile belt, when South America is restored in a fit described in the literature with respect to Africa. Both units yield stable, bipolar, high-temperature magnetizations that, on the basis of thermal demagnetization data, (super 40) Ar (super 39) Ar hornblende and biotite radiometric results, and Rb-Sr whole-rock and biotite radiometric results, appear to be indistinguishable in age at 1.9 to 2.0 Ga. With South America in this reconstruction, the corresponding paleomagnetic poles suggest that approximately 1,000 km of right-lateral motion has occurred between the West African and Guyana shields, probably along the Liberian Pan-African Belt. Furthermore, the Encrucijada and Harper poles are distinct from the 1.9 to 2.0-Ga paleomagnetic poles from the Kalahari Shield and tentatively suggest that relative motion has occurred between the Kalahari Shield and on the West African and Guyana shields since that time. Descriptors: absolute-age; Africa; amphibolites; Ar-Ar; Atlantic-region; continental-drift; crystalline-rocks; dates; Eastern Venezuela; Encrucijada Pluton; faults; granites; Guyana Shield; Harper; igneous rocks; Kalahar -Shield; La-Encrucijada; Liberia; metamorphic rocks; movement; orogeny; paleogeography; paleomagnetism; Pan African Orogeny; plate tectonics; plutonic rocks; pole positions; Precambrian; Proterozoic; Rb-Sr; South America; stratigraphy; strike slip faults; tectonophysics; upper Precambrian; Venezuela; West Africa; West African Shield; Solid earth geophysics. References: 48; illustrations, including 2 tables, sketch maps. ISSN: 0016-7606.

Order of the Holy Cross. 1942. "Northwestern Liberia." Hand drawn map. Order of the Holy Cross, Liberian Mission, Bolahun, Liberia. December 1942. 1 map, hand drawn. Scale 1:200,000 or "3.1567 miles to an inch." Notes: Bolahun- Vonjama, Zorzor area. Survey: compass time estimate method. 2 sheets, the south half of the map is apparently missing from the Library of Congress copy. Library of Congress, Geography & Map Division.

Orogun, P. S. 2003. "Plunder, Predation and Profiteering: The Political Economy of Armed Conflicts and Economic Violence in Modern Africa." Perspectives on Global Development and Technology, Volume 2, Number 2, 2003, pages 283-313(31). Abstract: This paper presents a comparative analytical study that is based on a political economy perspective concerning the effects of economic violence and the specter of predation-induced armed conflicts in modern African states. Although "blood diamonds," crude oil, "conflict timber," and illicit arms trafficking have engendered and exacerbated civil wars, cross-border raids, and protracted regional destabilization in Angola and the Democratic Republic of Congo, my primary focus is on the ongoing military debacle in Liberia and the recently concluded mayhem in Sierra Leone. The "resource curse" hypothesis will be utilized to examine and to illuminate the impact of economic pillaging, illicit arms trade, and predatory warlordism on the political instability and humanitarian atrocities in these two West African countries. A review of the internal regime types and the regional security relations within the sub-region will help to contextualize the recurrent trends and discernable systemic patterns that have been associated with these pillaging wars in the post-cold war era of Africa's international relations. In short, armed conflicts have weakened state capabilities, strained the financial resources of nongovernmental organizations and even raised provocative questions about the political will and sustaining capacities of the international community and regional security organizations to keep the peace and create conditions that are conducive to long-term, sustainable and viable political stability and economic development in the conflict-ridden and war-ravaged Sub-Saharan African States.

Ortmark, Åke, 1929- . 1977. Lamco!: en omstridd och betydelsefull svensk satsning I Afrika. Stockholm: Wahlström & Widstrand. 52 pages; 20 cm. Descriptors: Investments, Swedish – Liberia; Technical assistance, Swedish – Liberia; Liberia - Economic conditions. Named Corp: LAMCO. ISBN: 9146130284; LCCN: 78-385480 OCLC: 4591803.

"Overview of land-based sources and activities affecting the marine, coastal and associated freshwater environment in the West and Central African region." 1999. UNEP Reg. Seas Rep. Stud. no. 171, 117 pp. 1999. Descriptors: anthropogenic factors; international cooperation; environmental assessment; pollution effects; coastal zone management; urbanization; land use; marine environment; inland water environment; resource management; socioeconomic aspects; fishery resources; Africa, Northwest; Angola; Nigeria; Senegal. Abstract: This document provides a regional overview on land-based sources and activities affecting the marine, coastal and associated freshwater environment in the West and Central African region. It addresses natural conditions and processes, anthropogenic impacts and its socio-economic implications, including losses of cultural heritage sites. This overview also contains information on emerging and foreseeable problems in the region, proposing priorities for action including regional and international activities for cooperation. This document encompasses the following nations: Angola, Benin, Chad, Congo (Democratic Republic), Cote d'Ivoire, Gambia, Ghana, Guinea, Guinea-Bissau, Liberia, Mauritania, Nigeria, Sao Tome and Principe, Senegal and Togo. In summary, information received from various WACAF countries and obtained from a number of other sources shows that the major issues related to the marine, coastal and associated freshwater environment pollution in the region include: (a) The decline of water quality, due to land-based human activities, such as the introduction of sewage and waste water from industrial, domestic and agricultural run off as well as coastal urbanization; (b) Physical degradation and habitat modification; and (c) Fishery resources depletion and the loss of marine biodiversity. The socio-economic and cultural implications can be tremendous in terms of income reduction arising from a loss of fisheries stocks and catches, recreation and tourism amenities, increase of water treatment and coastal protection costs. Because of the lack of detailed scientific data on coastal, marine and freshwater environment in the WACAF region, a certain degree of uncertainty prevails in assessing the pollution load in general. There is an urgent need for a precise qualitative and quantitative assessment of the significant sources of land-based pollution in the region. Notes: 49 tables and 16 figures. ISBN: 9280718003.

Papadakis, Juan. 1966. Crop ecologic survey in West Africa (Liberia, Ivory Coast, Ghana, Togo, Dahomey, Nigeria) Prepared by J. Papadakis. [Rome] Food and Agriculture Organization of the United Nations, 1966. 2 volumes, color maps, tables 43 x 58 cm. Contents Notes: v. 1 Text.--v. 2. Atlas. Notes: Atlas in English and French. Descriptors: Crops and climate Africa, West Maps; Agriculture Africa, West Maps; Africa, West Climate; Food and Agriculture Organization of the United Nations. University of Wisconsin- American Geographical Society Collection. Call Number: S471.A395 P3

Paran-ko, I. S.; Adu, T. K. 1993. "Formatsii i stratigraficheskoye raschleneniye sistemy Birrimiy Leono-Liberiyskogo shchita Tsentral'noy Gany." Translated "Formations and stratigraphic differentiation of the Birrimian Leone-Liberian Shield, central Ghana." Otechestvennaya Geologiya. 1993; 4, Pages 36-40. 1993. Izdatel'stvo Nedra. Moscow, Russian Federation. 1993. Language: Russian. Descriptors: Africa; amphibolite facies; Birrimian; Central Ghana; epidote; epidote group; facies; Ghana; greenstone; Leone Shield; Liberia; Liberian Shield; metamorphic rocks; orthosilicates; Paleoproterozoic; Precambrian; Proterozoic; schists; Sierra Leone; silicates; sorosilicates; upper Precambrian; West Africa. Illustrations: References: 10; 2 tables. ISSN: 0869-7175.

Pearson, H. D., Capt. and Cox, E. W., Lieut. 1904. Map of the Sierra Leone-Liberia boundary. Royal Geographical Society (Great Britain). Publication: London: Royal Geographical Society. Description: 1 map. Descriptors: Boundaries- Sierra Leone- Maps; Boundaries- Liberia- Maps. Map Info: Scale 1:500,000; cylindrical projection. Notes: Date of information 1903. International boundary shown. Coverage: Eastern Sierra Leone and northwestern Liberia. Responsibility: Surveyed by Capt. H. D. Pearson & Lieut. E. W. Cox, Anglo-Liberian boundary commission. OCLC: 44822769.

Perlack, R. D.; Barron, W. F.; Samuels, G. and Rhinelander, R. E. 1985. Analysis of the Costs of Fuel Supply for Wood-Fired Electric Power Plants in Rural Liberia. Oak Ridge National Lab., TN, United States. Report number: ORNL6136. June 1985. 66p. Descriptors: Cost; Economic Analysis; Rural Areas. Liberia; wood fuels; wood fuel power plants. Abstract: In recent years the quality of rural electric services in Liberia has been declining and the future economic viability of these power stations is a growing concern. Each of the ten operating and each of the planned rural public power stations is designed to operate exclusively on gas oil (diesel fuel). Fuel expenditures by the Liberian Electricity Corporation (LEC) for the rural public stations represent a major and growing burden on the financially hard-pressed utility. Liberia has two potentially significant alternatives to oil-fired electric power for its up-country towns: small (1 to 5 MW) hydroelectric facilities, and wood-fired steam or gasifier plants (0.2 to 2 MW). Although small hydroelectric facilities appear viable for several locations, they cannot serve all locations and will require thermal back-up. The economics of supplying wood to a rural electric power plant or rural grid were evaluated under several scenarios involving: (1) different sources of the feedstock, and (2) differences in wood supply requirements for plants based on the use of steam or gasifier technology, and variation in the utilization level for such plants. With a few minor exceptions, wood energy supplies are plentiful throughout Liberia. Liberia has four different potential sources of wood fuel supply: the commercial cutting of retired rubber trees; the harvesting of secondary growth forest just prior to the land returning to temporary cultivation as part of a system of shifting agriculture; adding to the system of shifting agriculture the planting of fast-growing wood species and harvesting these trees when the land again is brought back under cultivation (generally after about five to seven years); and the establishment of commercial short-rotation wood energy plantations. Results indicate that the use of wood to fuel rural power stations is a viable economic option. NTIS Number: DE85015026XSP.

Perry, J. A. 1988. "Networking Urban Water Supplies in West Africa." Journal of the American Water Works Association JAWWA5 Vol. 80, No. 6, p 34-42, June 1988. 1 tab, 39 ref. University of Minnesota Agricultural Experiment Station Project 42-025. Abstract: The diversity among four West African water supply systems is characterized, and problems faced by managers of these kinds of utilities are outlined. After having interviewed municipal water system managers in the Ivory Coast, Ghana, Burkina Faso, and Liberia, the author describes physical facilities, monitoring and analysis operations, and the regulatory framework within which each system operates. Avenues of development that could improve the quality of regional water supplies include professional networks and twinning with utilities in the developed world. Such arrangements could provide a forum for exchange of resources among people addressing similar problems. Other productive avenues of change are development of information management systems for water quality data, improved performance indicators for in plant processes or community-level social services, and development of innovative management techniques. A significant role in developing such changes can be played by utility operators in developed countries. That role includes short- and long-term training to be provided in host countries or in schools and water supply systems in developed countries. Descriptors: Networks; Africa; urban areas; water supply; water management; regional planning; water distribution; Ivory Coast; Ghana; Burkina Faso; Liberia; water conveyance; management planning; public utility districts; information exchange; water quality

Peterson, James A. and Klemme, H. Douglas. 1986. "Geology and petroleum resources of northwestern Africa." In: AAPG annual convention with divisions; SEPM EMD DPA; technical program and abstracts. AAPG Bulletin. 70; 5, Pages 631-632. 1986. American Association of Petroleum Geologists. Tulsa, OK, United States. Non-USGS publications with USGS authors. Conference: American Association of Petroleum Geologists, 1986 annual meeting. Atlanta, GA, United States. June 15-18, 1986. Descriptors: Africa; Algeria; Bechar Basin; economic geology; energy sources; Essaouria Basin; Ivory Coast; Liberia; Mali; maturity; Mauritania; Morocco; North

Africa; offshore; Reggane Basin; Senegal; Sierra Leone; source-rocks; Taoudenni; Tarfaya Basin; Tindouf Basin; West Africa. ISSN: 0149-1423.

Petroconsultants S.A. 1970? Open acreage evaluation, Africa. Geneva: Petroconsultants S.A. Years: 1970-1979? Description: 5 v. (loose-leaf): ill., maps; 29 cm. Descriptors: Geology-Mauritania; Geology- Senegal; Geology- Gambia; Geology- Guinea-Bissau; Geology-Guinea; Geology- Sierra Leone; Geology- Liberia; Geology- Côte d'Ivoire; Geology- Ghana; Geology- Togo; Geology- Benin; Geology- Nigeria; Geology- Cameroon; Geology- Equatorial Guinea; Geology- Gabon; Geology- Congo (Brazzaville); Geology- Congo (Democratic Republic); Geology- Angola; Geology- South Africa; Geology- Botswana; Geology- Lesotho; Geology-Mozambique; Geology- Madagascar; Geology- Mauritius; Geology- Seychelles; Geology-Tanzania; Geology- Kenya; Geology- Uganda; Geology- Somalia; Geology- Sudan. OCLC: 14223168.

Phillips, C. 1972. "A Tourist Map of Liberia." Monrovia: The Bureau. 1 map, 25 x 39 cm. Scale not given. Notes: Produced by C. Phillips, Bureau of Tourism, MICAT. Includes points of interest for Montserrado County. LCCN: 85-690720.

"Physiography of Western Liberia." 1908. Scottish Geographical Magazine. Volume 24, pages 605-606. OCLC: 01604208.

Pomerene, Joel B. and Stewart, William E. 1967. Barite veins in the Gibi area of Liberia. Bulletin - Liberia, Geological Survey. Volume 1; 1967. Bureau of Natural Resources and Surveys. Monrovia, Liberia. Pages: 21. 1967. Abstract: Six deposits of high grade barites, mineralogy, spectrographic analyses, reserves. Descriptors: Africa; barite; economic geology; Gibi; Liberia; mineral deposits, genesis; reserves; sulfates; West Africa; Economic geology of nonmetal deposits. Illustrations, including maps. ISSN: 0459-2204.

Pomerene, Joel B. and Stewart, William E. 1967. "Barite veins in the Gibi area of Liberia." Bulletin of the Geological, Mining and Metallurgical Society of Liberia. 2; Pages 94. 1967. Geological, Mining and Metallurgical Society. Monrovia, Liberia. Descriptors: Africa; barite; economic geology; Gibi; Liberia; possibilities; sulfates; West Africa; Economic geology of nonmetal deposits. ISSN: 0367-4819.

Poorter, Lourens; Bongers, Frans and Sterck, Frank J. 2005. "Beyond the regeneration phase: differentiation of height-light trajectories among tropical tree species." The Journal of Ecology. Volume 93, no. 2 (April 2005) pages 256-67. Descriptors: Forest reproduction; Rain forests-Liberia; Tropical forest ecology- Liberia; Forest crown canopy- Light transmission. ISSN: 0022-0477.

Poorter L.; Bongers F.; Sterck F. J. and Wöll, H. 2003. "Architecture of 53 rain forest tree species differing in adult stature and shade tolerance." Ecology 84, no. 3 (01 March 2003) p. 602-608 Abstract: Tree architecture determines a tree's light capture, stability, and efficiency of crown growth. The hypothesis that light demand and adult stature of tree species within a community, independently of each other, determine species' architectural traits was tested by comparing 53 Liberian rain forest tree species. We evaluated whether species differed in their tree height, crown depth, and crown diameter, when compared at a standardized size of 15-cm diameter at breast height, and how their architecture changed early during ontogeny. Tree height was positively correlated with adult stature and light demand. By producing a relatively slender stem, large-stature species are able to rapidly reach their reproductive size, at a low cost for construction and support. Light-demanding species need a slender stem to be able to attain or maintain a position in the canopy. Both crown depth and crown diameter are negatively correlated with adult stature, but not with light demand. This is in contrast with the hypothesis that shade-tolerant species should have a shallow crown to reduce self-shading in a light-limited environment. Investing energy in height growth rather than lateral crown growth allows a rapid vertical stem extension, but crown diameter has to be sufficiently small to reduce the risk of mechanical failure. All architectural patterns were maintained during ontogeny. The key factors driving interspecific differences in tree architecture are the costs of height extension and mechanical stability. In general, light demand and adult stature represent independent axes of architectural differentiation, affecting tree architecture in different ways. References: 28. Descriptors: Plant ecology: general; shade tolerance. ISSN: 0012-9658.

Poorter, L.; Bongers, F.; van Rompaey, R. S. A. R. and De Klerk, M. 1996. "Regeneration of canopy tree species at five sites in West African moist forest." Forest Ecology and Management. Vol. 84, no. 1-3, pages 61-69. 1996. Descriptors: rain forests; regeneration; trees; population structure; Liberia; Ivory Coast. Abstract: Population structures were drawn for selected West African rain forest canopy tree species to evaluate whether regeneration was present. Regional variability was studied for five sites near the border between Liberia and Cote d'Ivoire. Population structures were highly variable. Three major types of population structures are recognised: a decrease in number of individuals with size, the typically inverse J-shaped curve indicating sufficient regeneration; an increase in number of individuals with size, indicating an absent or sparse regeneration; variable, consisting of strongly fluctuating patterns, in most cases many small individuals, no intermediate and many large ones. At a regional scale, most species show variable population structures. Population structures are static representations of a population composition at a certain moment in time. More long term information on population dynamics is needed to be able to interpret population structures correctly. To sustain the yield from tropical forest a minimum of juvenile trees of commercial species is required to compensate for logging of mature trees. The level of this minimum is hard to indicate without data on population dynamics. Regional variation in population structures warrants the use of local information as a basis for selection criteria of individual trees. ISSN: 0378-1127.

Porter, Philip Wayland. 1953? "Republic of Liberia Land Use." Monrovia. Publisher not given. Date not given. Scale 1:1,000,000. Hotine's rectified orthomorphic projection. 1 map, 57 x 58cm. Notes: Main source of air photo mosaics (1:20,000) and stereo-prints (1:40,000) covering 96 per cent of the country prepared by the Aero Service Corporation of Philadelphia, PA, United States. Nearly three quarters of the photographs were taken in the period February 1 to March 26, 1953. Includes reliability diagram. Insets of important cash crops: rubber, coffee, cocoa, palm kernel, kola and piassava. LCCN: 97-680008.

"Precious Metal Exploration in Eastern Liberia." 1978. United Nations Revolving Fund for Natural Resources Exploration. Report number LIB/NR/77/001-1. Monrovia, Liberia. 62 pages.

Premier essai de lexique du PAOB--First attempt for PAOB lexicon. 1980. In: Le Precambrien de l'Afrique de l'Ouest et ses correlations avec le Bresil Oriental; Bulletin d'information No. 4-5." Translated title: "The Precambrian of West Africa and its correlations with eastern Brasil; Newsletter No. 4-5." Yace, I. (leader). Pages 2-3. 1980. IGCP (International Geological Correlation Programme). Language: French; English. Descriptors: Africa; Birrimian; Brazil; cratons; cycles; Dahomeyides; evolution; Kibarian Orogeny; lexicons; Liberian Orogeny; Mauritanides; orogeny; Paleoproterozoic; Pan African Orogeny; Precambrian; Proterozoic; South America; stratigraphy; upper Precambrian; West Africa. Notes: IGCP Project No. 108/144.

Preuss, E. 1969. "Verschleppte Tektite in Liberia." Translated title: "Dispersed tektites in Liberia." Naturwissenschaften. 56; 10, Pages 512. 1969. Springer-Verlag. Berlin, Federal Republic of Germany. Language: German. Abstract: Source area in southeast Asia-Australia strewn field, fission track and K/Ar ages, chemical composition (see also this Bibliography Vol. 33, No. 1, 09 E69-01219). Descriptors: Africa; age; composition; dates; electron probe; find not authentic; fission track; geochronology; K-Ar; Liberia; petrology; regional; spectroscopy; tektites; West Africa; Igneous and metamorphic petrology. ISSN: 0028-1042.

Preuss, E. and Meyer von Freyhold, O. 1968. "Der erste Tektitfund in Liberia." Translated Title: "First discovery of tektites in Liberia." Naturwissenschaften. 55; 4, Pages 177-178. 1968. Springer-Verlag. Berlin, Federal Republic of Germany. Language: German. Abstract: Eastern Liberia, two tektites, distinctive characteristics, source in Bosumtwi crater (Ivory Coast) at a distance of 780 km. Descriptors: Africa; East; general-description; Liberia; petrology; tektites; West Africa. ISSN: 0028-1042.

Rabchevsky, George A. 198? "The mineral industry of Liberia." U. S. Bureau of Mines, United States. Pages: 7. 198?. Descriptors: Africa; economic geology; economics; Liberia; mineral resources; production; review; West Africa; Economic geology,-general,-economics. Notes: Preprint from Minerals yearbook, 1981, U. S. Bur. Mines.

RCA Victor Company, Ltd. 1970. "Geographical Layout of Liberian Telecommunications Systems." Montreal, Canada: RCA. 1 map, photocopy, 55 x 85 cm. Scale not given. Notes: "1086283." LCCN: gm71-1872.

Rebert, J. P. 1978. "Note on the hydrology of the West African continental shelf, from Mauritania to Guinea." French title: "Apercu sur l'hydrologie du plateau continental ouest-africain de la Mauritanie a la Guinee." Meeting of the Ad Hoc Working Group on West African Coastal Pelagic Fish from Mauritania to Liberia. Dakar (Senegal), 19 Jun 1978. FAO: Rome (Italy). Fisheries Dept. May 1979. Pages 87-92. Languages: French. Notes: Maps. Descriptors: Hydrology; Oceanology; Continental Shelves; Atlantic Ocean; West Africa. M Report No: FAO-FI--CECAF/ECAF-SERIES/78/10; FAO-FI--INT/72/074. Project: INT/72/074. Job No: N0952; N0321. Fiche No: 43519. Related item: 143512. Text in English, French. Acc. No: 143519. MFN: 143519. FAO Library.

Reconnaissance soils survey map of Liberia. 1968. Monrovia? No publisher given. Description: 1 map: photocopy; 29 x 45 cm. Descriptors: Soils- Liberia; Soils- Maps. Map Info: Scale not given. Notes: "Figure 4." Language: English. LCCN: 85-690702; OCLC: 12836268.

Reed, William Edward. 1951. "Reconnaissance Soil Survey of Liberia." US Department of Agriculture. Agriculture Information Bulletin 66. 107 pages. Washington, D.C.: U.S. Dept. of Agriculture, Office of Foreign Agricultural Relations and U.S. Dept. of State, Technical Cooperation Administration. OCLC: 5897702.

Rehfeldt, Warren R. 1967. "Geological investigations in the Bie Mountains area, Grand Cape Mount County, Republic of Liberia." Bulletin of the Geological, Mining and Metallurgical Society of Liberia. Volume 2; Pages 59-69. 1967. Geological, Mining and Metallurgical Society. Monrovia, Liberia. Abstract: Low grade banded iron formations, meteoric alterations, genesis, syncline structure, exploration and reserves, Precambrian metamorphosed sediments. Descriptors: Africa; Bie Mountains; economic geology; exploration; Grand Cape Mount County, Liberia; iron; iron deposits; Liberia; metals; metamorphic rocks; metasediments; mineral deposits, genesis; petrology; West Africa; Economic geology of ore deposits. Illustrations: sketch map. ISSN: 0367-4819.

Reilingh, Albert, 1919-. 2001. Liberia: window of time, 1948-1950. Victoria, B.C.: Trafford, Year: 2001. 105 p.: ill., map; 24 cm. ISBN: 1552128784. Descriptors: Geologists- Liberia-Biography; Mining engineers- Liberia- Biography; Géologues- Liberia- Biographies; Ingénieurs des mines- Liberia- Biographies. Notes: "A snap shot of the life of a mining engineer"--Cover. OCLC: 47786662.

"Report on Superficial Geological Survey of the Eastern Province and Basoa County." 1942. Mining World. June 27, 1942. Volume 142, page 362 et seq.

"Ressources en eau dans les zones rurales de l'Afrique du sud du Sahel." Translated title: "Water resources in the rural zones of Sahel of Southern Africa." 1984. Nature and Resources. 20; 2, Pages 27. 1984. UNESCO by Parthenon Publishing. Paris, France. Language: French. Descriptors: Africa; agriculture; Benin; East Africa; economic geology; Ethiopia; Ghana; hydrogeology; hydrology; Liberia; Sahel; Senegal; Sierra Leone; surveys; water resources; West Africa; Zambia; Hydrogeology. ISSN: 0028-0844.

Richards, J. G. 1982. "The GOL/Miners Pilot Project, Sirleaf Camp, Final Report, February 10 - July 17, 1982." Liberian Department of Mines unpublished un-numbered report, 21 pages. AMRS, Inc. See: http://www.africaminerals.com/

Richards, J. G. 1967. "Factors and problems in setting up an iron and steel industry in less developed areas." Geological, Mining & Metallurgical Society of Liberia. Bulletin. Volume 2, pages 75-77.

Richardson, Nathaniel R. Jr., 1980. "A Brief Outline of the Mineral Occurrences of Liberia." Liberian Geological Survey, un-numbered report July 1980, 43 pages. AMRS, Inc. See: http://www.africaminerals.com/

Richardson, Nathaniel R. 1973. Geologic interpretation of selected anomalies within the Monrovia Quadrangle, Liberia, using an integration of geophysical methods. Master's thesis, Michigan State

University. East Lansing, MI, United States. Pages: 63. Illustrations, maps; 29 cm. Descriptors: Africa; alkali metals; applications; Bouguer anomalies; cartography; crust; free air anomalies; geologic mapping; geologic maps; geophysical methods; geophysical surveys; gravity anomalies; gravity methods; intrusions; isostasy; Liberia; mafic composition; magnetic methods; maps; metals; Monrovia Quadrangle; petrology; potassium; radioactivity methods; remote sensing; structure; surveys; thickness; West Africa. Subjects: Geology- Monrovia, Liberia. Notes: Includes bibliographical references (leaves 60-61). 8 maps in a pocket. OCLC: 24861551.

"Les Richesse minerals du Liberia." 1947. La Chronique des Mines Colonials. Paris: Le Bureau d'études géologiques et minères coloniales. Language: French. 15e année, Numbers 134-135, 15 Août- 15 Septembre 1947. Pages 255-256. OCLC: 7879812.

"Roads." 1952? Liberian Department of Public Works and Utilities. Scale: 1:986,000. Notes: Photostat. Library of Congress.

Robb, James M.; Schlee, John and Behrendt, John Charles. 1973. "Bathymetry of the continental margin off Liberia, West Africa." Journal of Research of the U. S. Geological Survey. 1; 5, Pages 563-567. 1973. U. S. Geological Survey. Reston, VA, United States. Descriptors: acoustical methods; Africa; Atlantic Ocean; bathymetry; bottom features; continental margin; continental shelf; continental slope; geophysical methods; geophysical surveys; Liberia; marine; marine geology; offshore; surveys; valleys; West; West Africa; Applied geophysics. Illustrations, including sketch map. ISSN: 0091-374X.

Robson, P. 1982. "The Mano River Union." Journal of Modern African Studies. Volume 20, no. 4 (1982) pages 613-628. Abstract: Linking Liberia and Sierra Leone (since 1973) and (later) Guinea, the Union aims at economic and infrastructural integration. All three countries are resource-rich though Guinea, unlike the others, is francophone, lacks a Creole element, and espouses a rigidly State-controlled socialist system. To justify its continued existence, M.R.U. needs 'to deliver' and, for that, the member states must be prepared to compromise on joint development initiatives.

Rodier, J. 1963. "Bibliography of African Hydrology." Paris: United Nations Education, Science and Cultural Organization. 165 pages. English and French.

Rollinson, H. R. 1978. "Zonation of supracrustal relics in the Archaean of Sierra Leone, Liberia, Guinea and Ivory Coast." Nature (London). 272; 5632, Pages 440-442. 1978. Macmillan Journals. London, United Kingdom. Descriptors: Africa; Archean; correlation; cratons; Guinea; Ivory Coast; Liberia; lithostratigraphy; metamorphic belts; metamorphic rocks; Precambrian; relict materials; Sierra Leone; stratigraphy; supracrustal rocks; West Africa; West African Shield; zoning; Stratigraphy. References: 20; chart, geological sketch map. ISSN: 0028-0836.

Rosenblum, Sam. 1974. "Analyses and economic potential of monazite in Liberia." Journal of Research of the U. S. Geological Survey. 2; 6, Pages 689-692. 1974. U. S. Geological Survey. Reston, VA, United States. Descriptors: abundance; Africa; chemical analysis; data; economic geology; exploitation; feasibility studies; heavy minerals; Liberia; metals; mineral economics; monazite; phosphates; rare earths; West Africa; X-ray fluorescence; sketch map. ISSN: 0091-374X.

Rosenblum, Sam. 1969. "Chemical and spectrographic analyses of kyanite from Mt. Montro, Grand Bassa County, Liberia (Monrovia quadrangle J-24)." Washington: U.S. Geological Survey. 4 leaves: ill.; 27 cm. Series: Open-file report (Geological Survey (U.S.)); 1230. Notes: Referred to in press release dated May 21, 1969. Cooperative geologic program by the Liberian Geological Survey and the U.S. Geological Survey under the auspices of the Liberian Government and the Agency for International Development, U.S. Dept. of State. Descriptors: cyanite Liberia, Grand Bassa County; spectrum analysis. USGS Library.

Rosenblum, Sam. 1969. "Preliminary spectrographic analyses of monazite from western Liberia." [Menlo Park, Ca. s.n.]: US Geological Survey. Liberian Geological Survey. United States. Agency for International Development. 11 leaves: map; 27 cm. Series: Open-file report (Geological Survey (U.S.)); 1318. Notes: "Part of the cooperative geologic program currently being conducted by the Liberian Geological Survey and the U.S. Geological Survey under the auspices of the Liberian Government and the Agency for International Development, U.S. Dept. of State." Referred to in press release dated October 24, 1969. Bibliography: leaf 11. Descriptors: Monazite Liberia; spectrum analysis. USGS Library.

Rosenblum, Sam; Leo, G. W. and Srivastava, S. P. 2000. Methods and preliminary results of heavy-mineral studies in Liberia. U. S. Geological Survey, Denver, CO, United States; Liberian Geological Survey, Liberia. Open-File Report - U. S. Geological Survey. 2000. REPORT NUMBER: OF 00-0259. Descriptors: Africa; fluvial environment; gems; gold ores; heavy minerals; Liberia; metal ores; mineral assemblages; mineral composition; mineral exploration; placers; resources; sediments; separation; stream sediments; USGS; West Africa. References: 31; illustrations, including 2 tables, sketch maps. ISSN: 0196-1497. USGS Library: (200) R29o no. 2000-259. Online: http://onlinepubs.er.usgs.gov/djvu/OFR/2000/ofr_00_259.djvu

Rosenblum, S. and Srivastava, S. P. 1979. "The Bambuta phosphate deposit, Liberia; a reconnaissance report." Liberian Geological Survey. USGS Bulletin. Report number: B 1480. Pages (monograph): 26. U. S. Geological Survey, Reston, VA, United States. January 1, 1979. 26 pages: ill., maps; 24 cm. Descriptors: Africa; Bambuta; deposits; economic geology; exploration; Liberia; phosphate deposits; phosphates-Liberia; Precambrian; reserves; USGS; West Africa. Notes: Prepared in cooperation with the Liberian Geological Survey under the sponsorship of the Agency for International Development, U.S. Dept. of State. Bibliography: p. 25-26. Category Type: Economic geology of nonmetal deposits. ISSN: 8755-531X; LCCN: 79-14259; OCLC: 4883062.

Rosenblum, Sam and Srivastava, S. P. 1970. "Silica sand deposits in the Monrovia area, Liberia." Bulletin - Geological, Mining and Metallurgical Society of Liberia. 4; Pages 44-55. 1970. Notes: Vol. 4. Geological, Mining and Metallurgical Society. Monrovia, Liberia. Descriptors: Africa; Buchanan; clastic sediments; economic geology; genesis; Liberia; mineral deposits, genesis; Monrovia; sand; sediments; West Africa; Economic geology, general. Illustrations, including sketch map. ISSN: 0367-4819.

Rosenblum, Sam and Srivastava, S. P. 1969. "Silica sand deposits in the Monrovia area, Liberia." Washington: U.S. Geological Survey, [1970]. 12 leaves: ill.; 27 cm. Series: Open-file report

(Geological Survey (U.S.)); 1434. Notes: Bibliography: leaf 12. Descriptors: Sand,-Glass-Liberia-Monrovia. LCCN: 76-374144 //r83; OCLC: 1809483.

Ross, E. 1919. "Climate of Liberia and its effects on man." Geographical Review. Volume 7, no. 6, pages 387-402.

Rozario, Paul. 2003. Liberia. Milwaukee, Wis.: Gareth Stevens Pub. 2003. 96 pages: ill. (some color), color maps; 26 cm. Series: Countries of the world. Contents: Overview of Liberia: Geography- History- Government and the economy- People and lifestyle- Language and literature- Arts- Leisure and festivals- Food- Closer look at Liberia: American colonization society- Birds of Liberia- Dan masks- Diamonds; a controversial resource- Indigenous alphabets- Joseph Jenkins Roberts- Kpelle- Liberian folktales- Logging and the environment- Mining- Monrovia- River Horse- Rubber- Vai arts- William Tubman- Relations with North America. ISBN: 0836823664 (lib. bdg.); LCCN: 2003-45560. Abstract: Provides an overview of the geography, history, government, people, arts, foods, and other aspects of life in Liberia. Descriptors: Liberia- Juvenile literature. Notes: Includes bibliographical references (p. 94) and index. Material Type: Juvenile (no specific ages). OCLC: 51984958

Rupert, James. 1999. "Diamond Hunters Fuel Africa's Brutal Wars; In Sierra Leone, Mining Firms Trade Weapons and Money for Access to Gems." Washington Post (October 16, 1999):A, 1:2. Photograph; Map. Abstract: The key role of mining interests in the fighting was nothing new in Sierra Leone. The eight-year conflict that has shattered this country and brutalized its 5 million people has been fueled by foreigners' hunger for diamonds. Rival mining companies, security firms and mercenaries- from Africa, Europe, Israel and the former Soviet Union- have poured weapons, trainers and fighters into Sierra Leone, backing the government or the rebels in a bid to win cheap access to diamond fields. Charles Taylor, president of Sierra Leone's neighbor, Liberia, and his son, Charles Jr., have helped the RUF obtain foreign arms and military training, said African and Western military intelligence sources and Liberians. An American with military experience described watching at Liberia's main airport as members of one of the president's security forces supervised the unloading of two truckloads of automatic rifles and ammunition that he said were then sent to the Sierra Leonean border. A Liberian government spokesman denied that Taylor or his son had provided weapons to the RUF, or had interests in Sierra Leone's diamond trade. International diamond merchants and other sources say that by helping the RUF control Sierra Leone's diamond fields, Liberia can divert more Sierra Leonean diamonds through its territory on the way to world markets, reaping part of the profits. Descriptors: Civil war; Military sales; Mining industry; Diamonds. ISSN: 0190-8286.

Rusk, D. C.; Bennett, K. C. and Mohn, K. W. 2002. "Petroleum systems in Offshore Sierra Leone and Liberia." Conference Sponsor: Enterprise Oil, AGIP: Exploration and Production Division. Conference: 64th Annual Meeting of the European Association of Geoscientists and Engineers, Florence (Italy), 27-30 May 2002. (World Meeting Number 0005907). Notes: European Association of Geoscientists and Engineers, P.O. Box 593990 DB Houten, The Netherlands; URL: www.eage.nl. Poster Paper No. P204.

Saha, S. C. 1988. "Agriculture in Liberia during the nineteenth century: Americo- Liberians' contribution." Canadian Journal of African Studies 22, no. 2 (1988) pages 224-239. Abstract: The most pressing problem faced by Liberia during the nineteenth century was a stagnant economy. Shortage of funds inhibited government efforts to occupy hinterland territories effectively in order to prevent encroachment by European colonial powers and to implement a liberal or beneficent policy towards Liberia's indigenous population. In 1871 the treasury was empty, and President Edward James Roye was obliged to borrow 100,000 pounds from a British bank on quite onerous terms. When the loan became a political and economic problem and when cash-crop agriculture collapsed during the world trade depression from 1870 to 1900, the Republic plunged into debt obligations that it could not meet. By 1912 repeated defaults on loan repayments culminated with the establishment of an international receivership over all Liberian revenues.

Samuel, Eugenie. 2002. "Diamond wars: there's got to be a way to stop the trade that's funding terror. (This Week).(Kimberley Protocol seeks to outlaw conflict diamonds)." New Scientist. Volume 174, issue 2344. (May 25, 2002): p6(2). Abstract: Bloody conflicts being fought by rebels in Sierra Leone, Angola, Liberia and the Democratic Republic of Congo have one thing in common. They are financed by the illegal sale of diamonds from deposits in the areas they control. International efforts to halt this trade focus on the newly agreed Kimberley Protocol, which requires all diamonds to come with a certificate of origin. Though the scheme is not even in operation yet, New Scientist has learned that the human rights groups which brokered it are concerned that bogus certificates will make it almost useless. But perhaps the plan can be made to work: two teams of geologists are this week announcing possible ways of pinpointing where a particular diamond was mined. The success of the schemes will depend on governments being persuaded to fund a global database of the characteristics of stones from different regions.

Samuels, G. 1985. Liberian Energy Consumption and Sectoral Distribution for 1981. Performer: Oak Ridge National Lab., TN, United States. Funded by: Department of Energy, Washington, DC. Report number: ORNLTM9424; Contract number: AC0584OR21400; February 1985. 25p. Descriptors: Liberia; charcoal; commercial sector; data compilation; electric power; electric utilities; energy consumption; hydroelectric power; industry; mining; petroleum; residential sector; transportation sector; wood; energy conversion non propulsive conversion techniques; energy; energy use supply and demand; energy policies regulations and studies. Abstract: This report is one of a series of project papers providing background information for an assessment of energy options for Liberia, West Africa; it summarizes 1981 Liberian energy consumption data collected during 1982. Total Liberian primary energy consumption in 1981 was equivalent to 11,400,000 barrels of crude oil (BCOE) - 64% from wood, 31% from petroleum, and 5% from hydro. About 71% (8,100,000 BCOE) entered the domestic market. The difference represents exports (400,000 BCOE), refining losses (200,000 BCOE), and losses in converting wood to charcoal (2,600,000 BCOE). Of the 8,100,000 BCOE entering the domestic market, 58% was in the form of wood and charcoal, 35% petroleum products, and 7% hydro. Excluding wood and charcoal, electricity generation consumed 59% of the energy entering the domestic market. The three iron ore mining companies accounted for 60% of all electricity production; the Liberia Electricity Corporation for 35%, and private organizations and individuals for 5%. The mining operations (including electricity generation and transportation uses) consumed about 60% of all petroleum products. The transportation sector consumed 30% of all petroleum of which 85% was for road transport, 12% for the railroads owned and operated by the mining companies, and 3% for sea and air transport. Non transportation energy use in the industrial, commercial, government, and agriculture and forestry sectors is small. Together, these sectors account for less than 10% of the petroleum products consumed. Wood and charcoal were used almost entirely by the residential sector, which also consumed an additional 530,000 BCOE of other fuels. Over 90% of the 530,000 BCOE was for electricity and 290,000 (56%) was from petroleum. Over half of the petroleum (150,000 BCOE) was for generation at the mines for their associated communities. 8 references, 10 tables. (ERA citation 10:016396). NTIS Number: DE85007361XSP.

Samuels, G. 1985. Summary of Energy Planning Technical Support to the Government of Liberia. Performer: Oak Ridge National Lab., TN, United States. Funded by: Department of Energy, Washington, DC. Contract number: AC0584OR21400; Report number: ORNLTM9676. Report date: June 1985. 18p. Descriptors: buildings; cost; energy consumption; planning; wood fuels. energy balance; energy policy; Liberia; energy conservation; energy conversion non propulsive conversion techniques; energy policies regulations and studies; business and economics; foreign industry development and economics; energy use supply and demand. Abstract: Subsequent to a general assessment of energy options for Liberia, the principal activities of this program were: (1) an assessment of the economics of wood energy in Liberia; (2) a study of the potential for energy conservation in government buildings; (3) assistance in completing the 1982 Liberian energy balance; and (4) assistance in preparing the National Energy Plan. This report discusses the first three of these activities. A draft of the National Energy Plan was submitted in January 1985 to member agencies of the Liberian National Energy Committee for their review and comments. Liberia used the equivalent of 13.2 million barrels of crude oil in 1982- 67% from fuel wood, 4% from hydro, and 29% from imported petroleum. The wood was used almost entirely (approx. 99%) by the residential sector. Iron ore mining operations accounted for about 60% of domestic consumption of petroleum products. The transportation sector accounted for another 25%. The energy consumed by the agriculture and forestry sector was less than 2% of domestic consumption and was used primarily for operations of the large rubber plantations and timber concessions. Very little energy was used for food production. Significant energy savings in government buildings would require a major remodeling effort, including replacement of the louvered windows; extensive repairs to close large gaps around windows, air conditioners, and doors; and extensive caulking. The payback period from energy savings would be long. The assessment of the economics of wood energy indicates that wood can probably be delivered to a small rural power plant at costs that make this feedstock highly competitive for some and perhaps most of Liberia's rural electric stations. (ERA citation 10:048877). NTIS Number: DE85018174XSP.

Samuels, G.; Barron, W. F.; Barnes, R. W.; Hill, L .J. and Hobbs, B. F. 1985. Evaluation of the Liberian Petroleum Refining Company Operations: Crude Oil Refining Vs Product Importation. Performer: Oak Ridge National Lab., TN, United States. Funded by: Department of Energy, Washington, DC. Report Date: February 1985. 52p. Report number: ORNLTM9472; Contract number: AC0584OR21400. Descriptors: Liberia; petroleum refineries; cost; economic analysis; fuel oils; gas oils; imports; operation; prices; profits; sales; Liberia; chemistry, chemical engineering; propulsion and fuels; fuels; behavioral and social sciences, economics; energy fuels; energy- energy use, supply and demand; chemistry industrial chemistry and chemical process engineering; business and economics Foreign industry development and economics. Abstract: This report is one of a series of project papers providing background information for an assessment of energy options for Liberia, West Africa. It presents information on a controversial recommendation of the energy assessment - that the only refinery in the country be closed and refined products be imported for a savings of approximately $20 million per year. The report reviews refinery operations, discusses a number of related issues, and presents a detailed analysis of the economics of the refinery operations as of 1982. This analysis corroborates the initial estimate of savings to be gained from importing all refined products. 1 reference, 24 tables. (ERA citation 10:015686). NTIS Number: DE85007380XSP.

Sangmor, Samuel S. 1988. Geology, mineralization, and evolution of the Tortor Range, Bong County, Liberia. Description: xiii, 148 leaves, [2] leaves of plates: ill., maps 29 cm. Conference: Geological Society of America 1988 centennial celebration. Denver, CO, United States. October 31 to November 3, 1988. Descriptors: Africa; Archean; areal geology; Bong County; crustal shortening; evolution; geochemistry; gold ores; greenstone belts; hydrothermal alteration; Liberia; metal ores; metamorphic belts; metamorphic rocks; metamorphism; metasedimentary rocks; metasomatism; metavolcanic rocks; plate tectonics; Precambrian; retrograde metamorphism; rifting; stratigraphy; Tortor Range; West Africa. Subjects: Geology - Liberia. Geology, Structural. Physical geology. Notes: Two maps on folded leaves in pocket. Spine Geology of the Tortor Range, Liberia. Includes bibliographical references (leaves 128-138). Thesis (M.S.)--Wichita State University, Dept. of Geology, 1988. Other Titles: Geology of the Tortor Range, Liberia. OCLC: 18650464.

Sangmor, Sam S.1982. "The geology and mineral resources of the Gondoja area, Grand Cape Mount County, Liberia." Monrovia, Liberia: Liberia, Ministry of Lands, Mines and Energy, Dept. of Mineral Exploration & Research, Liberian Geological Survey. Description: 40 pages: maps; 28 cm. Descriptors: Geology- Liberia; Mines and mineral resources- Liberia. OCLC: 14696529.

Sangmor, Sam S. 1979. Report on the geology, gold, and other mineral resources of the Kokoya goldfields, Bong County, Liberia. Monrovia, Liberia: Republic of Liberia, Ministry of Lands, Mines & Energy, Liberian Geological Survey. 40 leaves, [1] leaf of plates: ill.; 28 cm. Notes: "September 1979." Bibliography: leaf 40. LC copy: page 4 bound upside-down. Descriptors: Geology; Liberia, Bong County; Gold ores; Liberia, Bong County; Mines and mineral resources- Liberia- Bong County. Notes: "September 1979." Bibliography: leaf 40. LCCN: 83-141867; OCLC: 10072000.

Sangmor, Samuel S. and Dorbor, J.K., 1982. "The Geology and Mineral Resources of the Gondoja Area, Grand Cape Mount County, Liberia." Liberia Geological Survey unpublished un-numbered report, 43 pages. AMRS, Inc. See: http://www.africaminerals.com/

Sangmor, Sam S.; Dorbor, Jenkins K.; Hoskins, Lynn; Mason, Jonathan A, Jr.; Murray, Gester and Pshorr, Peter. 1982. "The Mt. Dorthrow manganese mineralization." Liberian Geol. Survey Monrovia, Liberia. Pages: 53. 1982. Descriptors: Africa; airborne methods; chemically precipitated rocks; economic geology; geochemical methods; geophysical methods; geophysical surveys; Grand Gedet County; iron formations; Liberia; magnetic methods; manganese ores; metal ores; methods; mineral exploration; Mount Dorthrow; reserves; sedimentary processes; sedimentary rocks; surveys; West Africa. Notes: Project report of investigations, 1981. IGCP Project No. 108/144. References: 16; illustrations, including 4 tables, sketch maps.

Sangmor, Samuel S. and Nair, A. M. 1985. "Geology, Petrogenesis, Gold and Associated Mineralization in the Tortor Mountain Range, Bong County, Liberia." Liberia Geological Survey unpublished un-numbered report, 29 pages. AMRS, Inc. See: http://www.africaminerals.com/

Sass, J. H. and Behrendt, John Charles. 1980. "Heat flow from the Liberian Precambrian shield." Journal of Geophysical Research. 85; B6, Pages 3159-3162. 1980. American Geophysical Union. Washington, DC. Descriptors: Africa; chemically precipitated rocks; crust; geothermal gradient; heat flow; iron formations; Liberia; Precambrian; sedimentary rocks; shields; tectonophysics; temperature; West Africa; West African Shield. References: 26; illustrations, including tables, sketch map. ISSN: 0148-0227.

Schanze, E. 1982. "Joint Ventures zwischen Gaststaat und Investor im Erzbergbau." Translated title: "Joint ventures between the receiving country and the investor in the mining sector." Glueckauf (Essen). 118; 11, Pages 550-556. 1982. Verlag Glueckauf. Essen, Federal Republic of Germany. Language: German. Abstract: Legal problems that have risen from contracts between investors and the receiving countries of a mining project; two examples: the Bougainville copper mine in Papua New Guinea and the Bong iron mine in Liberia. Descriptors: Africa; Australasia; Bong Mine; Bougainville Mine; case studies; copper ores; corporate policy; economic geology; economics; global; iron ores; Liberia; metal ores; mines; news; Papua New Guinea; programs; public-policy; West Africa; Economic geology, general, economics. ISSN: 0340-7896.

Schlee, John S. 1976. "A selected look at Atlantic-type continental margins." Abstracts with Programs - Geological Society of America. 8; 2, Pages 261. 1976. Geological Society of America (GSA). Boulder, CO, United States. Conference: The Geological Society of America, Northeastern Section, 11th annual meeting; Southeastern Section, 25th annual meeting. Arlington, VA, United States. March 25-27, 1976. Descriptors: Africa; Atlantic type; Canada; Cenozoic; continental shelf; continental slope; evolution; faults; fracture zones; Indian Ocean; Liberia; Mesozoic; North America; Northern Hemisphere; northwest; oceanography; Red Sea; reefs; sedimentation; subsidence; tectonics; United States; West Africa; Solid earth geophysics. ISSN: 0016-7592.

Schlee, John S. 1972. Acoustic-reflection profiles, Liberian continental margin. Report number: PB- . Pages (monograph): 30. January 1, 1972. U.S. Geol. Surv., Reston, VA, United States. Abstract: Investigations carried out as part of a cooperative marine research program (International Decade of Ocean Exploration), studies of the geologic framework and resource potential of continental margins and small ocean basins. Annotation: Map. Source Note: USGS-GD-72-006. Descriptors: acoustical methods; Africa; Atlantic Ocean; continental shelf; geophysical methods; geophysical surveys; Liberia; oceanography; surveys; USGS; West Africa. Applied geophysics. Serial Report Available Only through NTIS.

Schlee, John. 1972. USGS IDOE 5. Geotimes. 17; 8, Pages 16-17. 1972. American Geological Institute. Alexandria, VA, United States. Abstract: Geophysical surveys, structure, continental margin off Liberia. Descriptors: Africa; Atlantic Ocean; continental shelf; continental slope; East; evolution; geophysical methods; geophysical surveys; gravity methods; Liberia; magnetic methods; ocean basins; oceanography; radioactivity methods; seismic methods; structure; surveys; West Africa; oceanography. Illustrations, including sketch map. ISSN: 0016-8556.

Schlee, John; Behrendt, John Charles and Robb, James M. 1974. "Shallow Structure and Stratigraphy of Liberian Continental Margin." AAPG Bulletin. 58; 4, Pages 708-728. 1974. American Association of Petroleum Geologists. Tulsa, OK, United States. Abstract: Rifting, fracture zones, Mesozoic-Tertiary sedimentation, geophysical surveys, dredged samples, Miocene foraminifera. Descriptors: Africa; Atlantic Ocean; biostratigraphy; carbonate rocks; Cenozoic; clastics; continental margin; continental shelf; continental slope; deposition; Foraminifera; fracture zones; geophysical methods; geophysical surveys; gravity methods; history; Invertebrates; Liberia; lithostratigraphy; magnetic methods; marine; marine geology; Mesozoic; microfossils; oceanography; offshore; onshore; Paleozoic; petroleum; Phanerozoic; Protista; reflection; rifting; sedimentary rocks; sedimentation; seismic methods; stratigraphy; structure; surveys; Tertiary; West Africa; Oceanography; sketch maps. ISSN: 0149-1423.

Schlee, John Stevens; Behrendt, John Charles and Robb, James M. 1973. "Shallow structure and stratigraphy of the Liberian continental margin." Washington: U.S. Geological Survey. 44 leaves: ill., maps; 28 cm. Open-file report (Geological Survey (U.S.)); 1905; Report number: OF 73-0249. Annotation: USGS, Woods Hole Oceanogr. Inst., Woods Hole, MA, United States. Notes: Referred to in press release dated September 7, 1973. Bibliography: leaves 37-41. Descriptors: Continental shelf, Liberia; Submarine geology- Liberia. USGS Library. ISSN: 0196-1497.

Schlee, John; Behrendt, John Charles and Robb, James M. 1972. "Shallow structure of the Liberian continental margin." Abstracts with Programs - Geological Society of America. 4; 7, Pages 655-656. 1972. Geological Society of America (GSA). Boulder, CO, United States. Descriptors: Africa; Atlantic Ocean; continental drift; continental shelf; continental slope; evolution; geophysical methods; geophysical surveys; Liberia; magnetic methods; ocean basins; ocean floors; oceanography; seismic methods; structure; surveys; West Africa; Oceanography. ISSN: 0016-7592.

Schmidt, M.; Borchers, R.; Fabian, P.; Flentje, G.; Matthews, W. A.; Szabo, A. and Lal, S. 1984. "Trace gas measurements during aircraft flights in the tropopause region over Europe and North Africa." Journal of Atmospheric Chemistry. Vol. 2, no. 2, pages 133-143. 1984. Descriptors: aircraft; gases; methane; carbon monoxide; troposphere; stratosphere; air sampling. Abstract: During aircraft flights in May 1981 from Munich (40 degree N) to north of the Spitsbergen Islands (82 degree N) and to Monrovia, Liberia (6 degree N), air samples were obtained in the altitude range of 8 to 11 km and during the ascents and descents near the airports. These samples have been analyzed for the trace gas mixing ratios of CH sub(4), CO and N sub(2)O. The results of these analyses are presented and discussed. The results provide new evidence of tropospheric-stratospheric exchange events in the vicinity of the subpolar and subtropical tropopause foldings and possibly show a case of transport of CO-enriched air in the upper troposphere above the North Atlantic Ocean. ISSN: 0167-7764.

Schultz, L.P. 1943. Freshwater fishes of Liberia. US National Museum. Proceedings. Volume 92, pages 301-303.

Schulze, Willi. 1965. Economic development and the growth of transportation in Liberia. Translation commissioned and edited by Peter K. Mitchell. [Freetown: Sierra Leone Association]. Description: Book. 29 p. maps. 28 cm. Series: Occasional paper (Sierra Leone Geographical

Association); no. 1. Notes: Translation of Liberia: Wirtschafts- und Verkehrsentwicklung, published in Geographisches Taschenbuch, 1964/65. Includes bibliographical references. Descriptors: transportation in Liberia; Liberia Economic conditions; Other: Geographisches Taschenbuch. University of Wisconsin- American Geographical Society Collection. Call Number: HC1075 .S38x 1965.

Schulze, W.O. 1967. "Der Eisenerzbergbau in Liberia." Translated title: "The iron ore mining industry in Liberia." Geographisches Rundschau. Volume 17, pages 443-454.

Schulze, W.O. 1964. "Infrastrukturfragen der Erzgebiete Liberia." Translated title: "Infrastructure questions of the ore areas of Liberia." Afrika Heute. November 1, 1964. Pages 274-281.

Schulze, W.O. 1964. "Early Iron Industry in the Putu Range in Liberia." Liberia University Journal. Volume 4, no. 2, pages 29-35.

SEAGAP exploration study, west Africa: Area 2, Senegal to Benin. 1977. Hispanoil. Exploration Division.; SEAGAP. Madrid: Hispanoil, Exploration Division; SEAGAP. 2 volumes: ill., maps; 29 cm. Contents: v. 1, case containing text plus assorted maps; v. 2, loose-leaf notebook containing seismic sections. Descriptors: Geology, Stratigraphic- Africa, West; Geology, Stratigraphic- Guinea; Geology, Stratigraphic- Sierra Leone; Geology, Stratigraphic- Liberia; Geology, Stratigraphic – Côte d'Ivoire. Cover title. "EWGS-Leg 2" "ES 0875". OCLC: 14223081.

Seagull Exploration Corporation. 1973. Petroleum exploration guide, African Atlantic basins: of Morocco, Spanish Sahara, Mauritania, Senegal, Guineas, Sierra Leone, Liberia, Ivory Coast, Ghana, Togo, Dahomey, Nigeria, Cameroun, Rio Muni, Gabon, Congo, Cabinda, Zaire, Angola, South Africa, Southwest Africa and general outline of West African interior basins. London: Seagull Exploration. Description: 1 case: ill., maps; 25 cm x 26 cm. x 34 cm. Descriptors: petroleum geology- western Sahara; petroleum geology- Equatorial Guinea; petroleum geology- Sierra Leone; petroleum geology- Liberia; petroleum geology- Côte d'Ivoire; petroleum geology- Africa, West. Notes: Case includes text, assorted maps and illustrations. "September, 1973". "Geology, plate tectonics, basin parameters, reserves, potential, prospects, ratings, regulations, exploration activity, political, economics, statistics, weather, recommendations." OCLC: 14269226

"Section of southern Montserrado County (Robertsfield-Oau road guide)." 1979. Liberian Cartographic Service. [Monrovia]: The Service. 1 map: photocopy; 53 x 125 cm. Subjects: Montserrado County (Liberia)- Road maps. Map Info: Scale 1:50,000. Notes: "LCS-98/79." Other Titles: Southern Montserrado County; Robertsfield-Oau road guide. LCCN: 85-690709; OCLC: 12836332.

Seitz, J. F. 1977. "Geologic map of the Voinjama Quadrangle, Liberia." Miscellaneous Investigations Series Map. Date: January 1, 1977. Report number: I-0771-D. U. S. Geological Survey, Reston, VA, United States. Map Type: geologic map. Color map; 44 x 59 cm. on sheet 74 x 104 cm. folded in envelope 30 x 24 cm. Map Scale 1:250,000 (1 inch = about 4 miles). Sheet 29 by 41 inches. Descriptors: Africa; areal geology; geologic maps; Liberia; maps; USGS; Voinjama Quadrangle; West Africa. Notes: "Hotines rectified skew orthomorphic projection and rectangular coordinates." "Prepared by the U.S. Geological Survey and the Liberian Geological Survey under the joint sponsorship of the Government of Liberia and the Agency for International Development, U.S. Department of State." Includes bibliography. ISSN: 0160-0753.

Seitz, J. F. 1977. "Geologic map of the Zorzor Quadrangle, Liberia." Miscellaneous Investigations Series Map. Report number: I-0773-D. January 1, 1977. U. S. Geological Survey, Reston, VA, United States. Bibliographic level: Monograph- color map; 44 x 44 cm. on sheet 74 x 107 cm. folded in envelope 30 x 24 cm. Map scale: 1:250,000 (1 inch = about 4 miles). Sheet 29 by 42 inches. Descriptors: Africa; areal geology; geologic maps; Liberia; maps; USGS; West Africa; Zorzor Quadrangle. Notes: "Hotines rectified skew orthomorphic projection and rectangular coordinates." "Prepared by the U.S. Geological Survey and the Liberian Geological Survey under the joint sponsorship of the Government of Liberia and the Agency for International Development, U.S. Department of State." Includes bibliography. ISSN: 0160-0753.

Seitz, J. F. 1974. "Geology of the Voinjama Quadrangle, Liberia." Open-File Report. Report number: OF 74-0301. Pages (monograph): 16 pages (1 sheet). January 1, 1974. U. S. Geological Survey, Reston, VA, United States. Annotation: 1 sheet, scale 1:250,000 (1 inch = about 4 miles). Descriptors: Africa; areal geology; Liberia; maps; USGS; Voinjama Quadrangle; West Africa. Notes: Map in pocket. Liberian investigations, Project report (IR) LI-68B. Prepared under the auspices of the Govt. of Liberia and the Agency for International Development. Letter of transmittal dated December 13, 1974. Bibliography: leaves 15-16. ISSN: 0196-1497.

Seitz, J. F. 1974. "Geology of the Zorzor Quadrangle, Liberia." Open-File Report. Report number: OF 74-0303. Pages (monograph): 14 pages (1 sheet). January 1, 1974. U. S. Geological Survey, Reston, VA, United States. Annotation: 1 sheet, scale 1:250,000 (1 inch = about 4 miles). Descriptors: Africa; areal geology; Liberia; USGS; West Africa; Zorzor Quadrangle. Notes: Map in pocket. Liberian investigations, Project report (IR) LI-69B. Prepared under auspices of Govt. of Liberia and Agency for International Development. Letter of transmittal dated December 9, 1974. Bibliography: leaves 13-14. ISSN: 0196-1497.

Seliverstov, Yu. P. 1978. "Surficial deposits and topography in an area of recent laterite formation in the southwestern Sahara platform." Lithology and Mineral Resources. 13; 2, Pages 164-174. 1978. Consultants Bureau. New York, NY, United States. Descriptors: Africa; applications; bauxite; Cenozoic; economic geology; exploration; geomorphological methods; geomorphology; Guinea; Guinea Bissau; Ivory Coast; landform evolution; laterites; Liberia; Mali; Mesozoic; mineral exploration; occurrence; ore deposits; Sahara; Sierra Leone; soil group; soils; surveys; West Africa; Economic geology of ore deposits; References: 17; illustrations, including sects., block diag., sketch map. ISSN: 0024-4902.

Sentab (Firm). 1962. Harbour at L. Buchanan plan: design Sept. 1961. Stockholm: Sentab, Edition: Rev. ... March 17, 1962. Description: 1 map: photocopy; 71 x 98 cm. Descriptors: Harbors- Liberia- Buchanan- Maps. Scale 1:2,000. Notes: Depths shown by contours. At head of LAMCO Joint Venture Nimba Mining Project. "Date: Jan. 30, 1961." Oriented with north toward the lower right. Includes table of coordinates. "A3-1-41." LCCN: 97-680005; OCLC: 36337392.

Sentab Consulting Engineers. 1961. "Nimba Mining Project." Scale 1:20,000. Notes: Blue line print. Descriptor: Liberia- Mines and Minerals. Library of Congress, Geography & Map Division.

Shannon, Eugene H. 1985. "The Precambrian of Liberia and its associated mineralisation--Le Precambrien du Liberia et ses mineralisations associees." In: 13th colloquium of African geology; abstracts --13 (super eú colloque de geologie Africaine; resumes. Bowden, Peter (editor) et al. Publication Occasionnelle - Centre International Pour la Formation et les Echanges Geologiques = Occasional Publication - International Center for Training and Exchanges in the Geosciences. 3; Pages 62-63. 1985. Centre International pour la Formation et les Echanges Geologiques (CIFEG). Paris, France. Conference: 13th colloquium of African geology-13 (super eú colloque de geologie Africaine. St. Andrews, United Kingdom. September 10-13, 1985. Language: English; French. Descriptors: Africa; economic geology; gneisses; iron ores; Liberia; metal ores; metamorphic rocks; metamorphism; Precambrian; regional metamorphism; schists; stratigraphy; West Africa; West African Shield; western Liberia; Stratigraphy; Economic geology, geology of ore deposits. ISSN: 0769-0541.

Shannon, Eugene H. 1982. "Brief review of the age provinces of Liberia and their correlations with South America based on (super 40) Ar (super 39) Ar dating and paleomagnetism." Liberian Geol. Surv., Liberia. Pages: 2. 1982. Report: IGCP (International Geological Correlation Programme). Descriptors: absolute age; Africa; Archean; dates; Eburnean Province; Guyana Shield; K-Ar; Liberia; Liberian Province; metamorphic rocks; paleomagnetism; Pan African Province; Precambrian; Rb-Sr; South America; stratigraphy; Venezuela; West Africa; Stratigraphy. Illustrations: sketch map. Notes: IGCP Project No. 108/144.

Shannon, Eugene H., et al. 1987. "Gold and Diamond Resources of the Lofa River Basin, Liberia." An unpublished and un-numbered report of the Liberia Geological Survey, marked as prepared "by an independent geologist." This may have been a report prepared by a consultant for a company thinking of investing in Liberia or it may have been prepared by the Geological Survey. Eugene Shannon was the Director of the Liberian Geological Survey at this time. AMRS, Inc. See: http://www.africaminerals.com/

Shannon, Eugene H. and Sangmor, S. S. 1983. "Geology for development; mineral resources and exploration potential of Liberia." In: Geology for development; mineral resources and exploration potential of Africa. Kogbe, Cornelius A. (editor). Journal of African Earth Sciences. 1; 3-4, Pages 369. 1983. Pergamon. London-New York, International. Conference: Sixth general conference on African geology; Geology for development; mineral resources and exploration potential of Africa. Nairobi, Kenya. December 11-19, 1982. Descriptors: Africa; economic geology; economics; Gola Forest Reserve; Kpo-Range; Liberia; mineral resources; production; West Africa. ISSN: 0731-7247.

Shannon, Eugene H. and Sangmor, Sam S. 1982. "Geology for development: a survey of the principal mineral resources of Liberia and their exploration potential." Liberian Geological Survey. [Monrovia, Liberia]: Liberian Geological Survey. 24 leaves: ill., maps; 28 cm. Descriptors: Mines and mineral resources- Liberia; Geology- Liberia. Notes: "Presented at the 7th African Geological Conference (GSA Biannual Conference), Nairobi, Kenya, December 13-19, 1982." OCLC: 12014546.

Sherman, Arthur. 1954. Liberia, report on the Bureau of Mines & Geology for the period October, 1952-September, 1953. Monrovia: Hydrographic Survey. Abstract: Includes data on the results of field work, geophysical and mineral investigations, and mining activity (iron, gold, diamonds). Descriptors: Africa; Bureau of Mines and Geology; Liberia; surveys, reports; West Africa. hydrographic survey, Freeport of Monrovia, Liberia. Liberian Cartographic Service. Description: 1 map: photocopy; 69 x 81 cm. Subjects: Harbors- Liberia- Monrovia- Maps. Scale 1:5,000. Notes: "September 1954." Depths shown by contours and soundings. Blue line print. Other Titles: Freeport of Monrovia, Liberia. LCCN: 97-680002; OCLC: 36337373.

Sherman, Arthur. 1951. "Possibilities and costs of methods of mineral discovery in Liberia." United Nations Sci. Conf., Pr.. 2; Pages 75-78. 1951. Descriptors: Africa; Liberia; Liberia, Bomi Hills-Bambuta area, phosphate mineralization; mineral discovery; mineral resources; West Africa. illustrations; discussion, pages 97-103.

Sherman, Arthur. 1948. "Map of Liberia." By Mr. Arthur Sherman, Mining Engineer. 1947, printed 1948. Scale: 1:534,000. Topographic map, 1:600,000. 1 sheet in color. Shows mountains as "wooly caterpillars." Notes: Institute of Geographical Exploration, Harvard University. "Work done during reconnaissance-mineralogic surveys from 1941-1944 for the Liberian Government, Bureau of Mines. Exploratory surveys with sketching board, controlled by prismatic compass and timed distances together with pedometer. Boundary line drawn from (1) British map of Sierra Leone showing tribal boundaries agreeing with map in 1929 British Dominion's Office and Colonial List. (2) Boundary map between Sierra Leone and Liberia, the British-Liberian Boundary Commission, 1913-1914 (Treaty Series 1917, No. 10 6 9027). (3) Republic of Liberia Boundary Commission, Boundary Surveys, Ben F. Powell, Chief Engineer (Sheets 1 to 7, inclusive) 1929. Coast from British Admiralty Charts and aerial photos. Only towns of a more or less permanent character have been shown. Areas indicated as thinly populated have a number of scattered towns of little importance. Names of towns and streams are listed as pronounced by the particular tribe. For example, Senbul correct Mano River name for misnamed Sanokwelli; Kpelli Lombaba correct Grande name instead of Lomboba (Lorma). Capital letters represent chiefdoms example Gibi Grande Bopolu. Held by: Library of Congress; USGS Library.

Siebel, Claus N. A. 2000. Die Eisenerzbergbau-Gesellschaft Bong Mining Company in Liberia, West Afrika 1955-1990. Düsseldorf: Stahleisen, 2000. ix, 230 p.: ill. (chiefly color), plans; 24 cm. Language: German. ISBN: 351400658X. Subjects: Corporations, German- Liberia. Named Corp: Bong Mining Company. Notes: Includes bibliographical references (p. 197-206). Other Titles: Bong Mining Company Liberia. OCLC: 46890317.

Da Silva, A. 1977. "As orogeneses assinaladas na regiao do Cariango (Angola) e consideracoes acerca do Sistema do Oendolongo e da Serie do Sansicua (Sistema do Congo Ocidental)." Translated title: "The marked orogenesis in the Cariango region (Angola) and the Oendolongo System and the Sansicua Series, Western Congo System." Garcia de Orta, Serie de Geologia. 2; 1, Pages 45-64. 1977. Junta de Investigacoes Cientificas do Ultramar. Lisbon, Portugal. Language: Portuguese; Summary Language: English. Descriptors: Africa; Angola; Cariango; Central Africa; gneisses; igneous activity; Limpopo, Liberia, Orogeny; lithostratigraphy; metaigneous rocks; metamorphic rocks; metamorphism; metasedimentary rocks; migmatites; Mussende Haco Group; orogeny; orthogneiss; paragneiss; petrology; processes; quartzites; Quitubia Group; schists;

sedimentation; structural geology; tectonics; Western Congo System. References: 80; illustrations, including geol. sketch map. ISSN: 0378-1240.

Skillings, D. N., Jr. 1968. "Bong Mining Co. Shipping Iron Ore Concentrate at Record Rate from Expanded Plant in Liberia, West Africa." Skillings Mining Review. Volume 57, no. 18, pages 1, 8-14. ISSN: 0037-6329; OCLC: 01765605.

Skillings Mining Review. 1964. "Bong Range- a new Iron Mining Project in Liberia, West Africa." Volume 53, no. 10, March 7, 1964, pages 4-5. ISSN: 0037-6329; OCLC: 01765605.

Skillings Mining Review. 1965. "Bong Project contributes to Liberia's Booming Iron Industry." Volume 54, no. 29, pages 4-5, 29. ISSN: 0037-6329; OCLC: 01765605.

Smith, H. F. 1931. "Resume of Report on Sanitation and Yellow Fever Control in Liberia." Washington: US Government Printing Office. "Number 1481." 7 pages.

"Soil Survey of Central Agricultural Experiment Station, Suakoko, Liberia." Notes: Publ. in cooperation with Liberia, Ministry of Agriculture, and the US Department of State, Agency for International Development and the USDA, Soil Conservation Service. Washington, D.C., United States. Pages: 31, [7] folded leaves of plates: illustrations; 29 cm. 1977. Subjects: Soils- Liberia-Suakoko- Maps. Soil surveys- Liberia- Suakoko. Descriptors: Africa; classification; genesis; Liberia; management; maps; soil assemblages; soil surveys; soils; soils maps; Suakoko; surveys; utilization; West Africa. Map scale: 1:20,000. Named Corp: Central Agricultural Experiment Station, Suakoko, Liberia. Notes: Cover title. Bibliography: p. 14. References: 7; illustrations, including tables. LCCN: 77-604152; OCLC: 3708141.

Somah, Syrulwa Lawson. 1994. "Historical settlement of Liberia and its environmental impact" 211 pages; [Ph.D. dissertation].United States -- Ohio: The Union Institute; 1994.
Abstract (Document Summary): This research argues that the historical settlement of liberated Americans of African descent had a destabilizing impact on the geography, politics, social and economic structure, environment and culture of Liberia. The research delves into the principal health problems (Plasmodium falciparum and Plasmodium malariae) which have historical roots, and also into those industrial activities which threaten the country's environment. Corruption, economic cataclysm, environmental degradation, and the absence of a regulatory ministry are scrutinized in the research. 90% of Liberia's population migrated into the hinterlands and never exercised full control over their nation's destiny. Now they have taken up a fight to forge a national policy which will lift them from the bottom of Liberia's political and socio-economic structure and reunite their nation. The causes of the 1980 military coup d'etat and 1989 civil war totaling approximately 200,000 casualties, reflect the schism in national vision. Case studies (qualitative, quantitative, comparative) calculate the overall impact of the settlement policy on Liberia's environment. This research examines the factors and co-factors hindering Liberia's national development and the results of the cultural clash between liberated Americans of African descent and Indigenous Liberians from the 1820s to the twentieth century. Pervasive arguments detail how cultural "non-fusion" created environmental, political, social and economic barriers. This document sets forth to prove that if they are not corrected, the current status of Liberia's social structures will

be exacerbated by the negative impact of social exploitation and separation presently evident in its society. DAI-B 55/10, p. 4311, April 1995.

Soubrier, J. 1936. "Au Libéria par la forêt tropicale." Translated title: "With Liberia through the tropical forest." La Géographie. Juillet 1936, Volume 66 (1), pages 1-18.

Srivastava, S. P. 1971. "Chrysoberyl in Liberia." Geological Survey of Liberia. Bulletin. Number 3. ISSN: 0459-2204; OCLC: 5356681.

Srivastava, S. P. 1970. "Mineralogy of phosphate rock from Bambuta." Bulletin - Geological, Mining and Metallurgical Society of Liberia. 4; Pages 26-32. 1970. Notes: Vol. 4. Geological, Mining and Metallurgical Society. Monrovia, Liberia. Descriptors: Africa; Bambuta; Bomi Hills; Liberia; mineralogy; minerals; occurrence; phosphates; properties; West Africa. Illustrations. ISSN: 0367-4819.

Srivastava, S.P. 1961. "The General Geology and Mineral Resources of the Republic of Liberia." Liberia Geological Survey unpublished un-numbered report, 38 pages. AMRS, Inc. See: http://www.africaminerals.com/

Stanin, S. Anthony and Cooper, Bismark R. 1968. "The Mt. Montro Kyanite deposit, Grand Bassa County, Liberia." Washington: U.S. Geological Survey. 33 leaves: ill., maps (1 folded); 28 cm. Series: Open-file report (Geological Survey (U.S.)); 1022. Notes: Prepared cooperatively by the Liberian Geological Survey and the U.S. Geological Survey as part of the Geological Exploration and Resources Appraisal Program sponsored by the Government of Liberia and the U.S. Agency for International Development. Referred to in press release dated April 22, 1968. United States Agency for International Development. Bibliography: leaves 32-33. Descriptors: Cyanite- Liberia, Grand Bassa County. USGS Library.

Stanin, S. Anthony and Cooper, Bismarck R. 1967. "The Mount Montro kyanite deposit Grand Bassa County, Liberia." Bulletin of the Geological, Mining and Metallurgical Society of Liberia. 2; Pages 97-98. 1967. Geological, Mining and Metallurgical Society. Monrovia, Liberia. Descriptors: Africa; ceramic-materials; economic geology; Grand-Bassa-County-Liberia; kyanite; Liberia; nesosilicates; orthosilicates; possibilities; reserves; silicates; West Africa; Economic geology of nonmetal deposits. ISSN: 0367-4819.

Stanin, S. Anthony and Cooper, Bismarck. 1967. Liberian Geological Survey. Field descriptions of rocks from measured sections, Mt. Montro kyanite deposit, Grand Bassa County, Liberia. Other Titles: Mt. Montro kyanite deposit, Grand Bassa County, Liberia. Monrovia, Liberia. 13 leaves: illustrations, maps (some folded); 35 cm. Series: Open-file report (Geological Survey (U.S.)); 1074. Also: Liberian Investigation (IR LI-15). Notes: Some of illustrative matter in pocket. Prepared in cooperation with Liberia Geological Survey. Referred to in press release dated July 19, 1968. Descriptors: Cyanite- Liberia, Grand Bassa County. USGS LIBRARY: (200) R29o no.1074.

Stanley, William. 2005. "Background to the Liberia and Sierra Leone implosions." GeoJournal, Volume 61, Number 1, January 2005, pp. 69-78(10). Abstract: Liberia and Sierra Leone are tragic examples of what happens when central authority collapses and warlords emerge as *de-facto* rulers over large portions of the national territory. Horrors inflicted on non-combatants and the well publicized trading in 'conflict diamonds' served to focus world attention on these two small countries sharing a common border. Both countries have experienced mixed success with outside military intervention for peace keeping and nation building purposes. What has happened is all the more distressing when one considers each country's prospects at birth under the political and economic aegis of arguably two of the most powerful and enlightened countries of the time, Great Britain in the case of Sierra Leone and the United States in Liberia. Descriptors: Americo-Liberian; collapse of central authority; corruption; diamonds; Krio (Creole); peacekeeping; levantine traders.

Stanley, William Richard. 2000. "Air transport in Liberia." Journal of Transport History 21, no.2 (2000) p. 191-208. ISSN: 0022-5266. Abstract: In World War II Liberia became one of the first West African bridgeheads of a civil-military air corridor from the United States via Brazil. Planes and supplies funneled along this route were instrumental in the British victory at El Alamein and, later, in the trans-Himalaya re-supply of China. This early start in land-based aviation was due to secret agreements between the United States and Britain and to the American decision quietly to subsidize Pan American World Airways' Caribbean and South American facilities. Pan Am already had a flying-boat service to Africa via Liberia and was charged with the construction of a pivotal inland airfield. The airfield would have been closer to Monrovia but for the strategic importance of Firestone's rubber plantations. After the war these linkages contributed to Liberia having one of the continent's first international airlines. Domestic air transport played a seminal role in improving communications between the capital and the settler communities along the coast. By the 1950s it was opening up areas of the isolated interior to small-scale operators exploiting alluvial diamond deposits. Later, air transport provided the initial connections for mining the country's iron ore and exploiting the tropical forest.

Stanley, William Richard. 1966. "Changing patterns of Transportation Development in Liberia." 218 pages; [Ph.D. dissertation].United States, Pennsylvania: University of Pittsburgh; 1966.

Stanley Engineering Company of Africa. 1964. Republic of Liberia, Mt. Coffee Hydroelectric Project, William V.S. Tubman Station, area and facilities map, transmission lines, contract 106. Stanley Engineering Company of Africa, consulting engineers-architects. Monrovia; Washington, D.C.: The Company. Cartographic Material. Description: 1 map: photocopy; 66 x 103 cm. Scale [ca. 1:250,000]. Notes: Covers Monrovia region. "Date: 13 March 1964." Relief shown by hachures. "Drawn: L.A. Satterthwaite." Oriented with north toward the upper left. Includes coverage map. "No. 3092-1." Subjects: Hydroelectric power plants- Liberia- Monrovia Region Maps. LCCN: 97-680001.

Starr, Frederick. 1918. "Map of Liberia." Scale 1:2,150,000. Library of Congress, Geography & Map Division.

Steck, H. E. 1965. "Die Aufbereitung in Bong Range." Deutscher hebe- u. forder-teknik. Volume 11 (5), pages 42-46.

Stewart, W. E. and Kromah, F. 1987. "Hydrocarbon generation potential of Monrovia, Roberts Bassa and Cestos basins of Liberia." In: Petroleum geochemistry and exploration in the Afro-Asian region. Kumar, Ruby K. (editor) et al. Pages 133-134. 1987. A. A. Balkema Publishers:

Rotterdam, the Netherlands. Conference: First international conference on Petroleum geochemistry and exploration in the Afro-Asian region. Dehra Dun, India. November 25-27, 1985. Descriptors: Africa; Cestos Basin; Cretaceous; economic geology; geochemical methods; Liberia; Lower Cretaceous; Mesozoic; Monrovia Basin; natural gas; offshore; petroleum; petroleum exploration; resources; Roberts Bassa Basin; West Africa. Illustrations: sketch map. ISBN: 90-6191-791-3.

Stewart, William E.; Freeman, William Amara and Dinkins, Daniel S. B. 1969. Field investigation of lot no. 3, near Bambuta, Lofa County: portion of Liberia Mining Company concession area and mosaic block G-16. Monrovia [Liberia]: Liberian Geological Survey, Bureau of Natural Resources & Surveys. Description: 9 leaves; 28 cm. Series: M[emorandum] r[eport] - Liberian Geological Survey; 53. Descriptor: Iron ores- Liberia, Lofa County. OCLC: 1812459.

Stibig, H. J. and Baltaxe, R. 1993. "Use of NOAA remote sensing data for assessment of the forest area of Liberia." RSC Series (FAO), no. 66. Rome (Italy) 21 pages, color illustrations Notes: Summaries (En, Es, Fr). Descriptors: forestry- general aspects; tropical forests; remote sensing; forest surveys; data collection; data analysis; methods; classification. FAO Library.

Stipp, H. E. 1976. "The mineral industry of Liberia." Minerals Yearbook. 1973, Vol. 3; Pages 581-586. 1976. U. S. Bureau of Mines. Washington, DC, United States. Descriptors: 1973; Africa; economic geology; economics; industry; Liberia; metals; mineral resources; nonmetals; petroleum; production; West Africa; Economic geology, general. Illustrations: tables. ISSN: 0076-8952.

Stipp, Henry E. 1974. "The Mineral Industry of Liberia." In: Minerals Yearbook 1972; Volume III, area reports; international (edited by Albert E. Schreck). Pages 541-546. 1974. U. S. Bur. Mines, Washington, D. C. Descriptors: 1972; Africa; data; economic geology; economics; Liberia; mineral resources; production; West Africa. Held by the USGS Library.

Stobernack, Just. 1970. "Metamorphose der Glimmerschiefer und Gneise in der westlichen Bong Range Liberia." Translated title: "Metamorphism of the mica schists and gneisses in the western Bong range, Liberia." In: Montangeologische Untersuchungen an Lagerstaetten des Eisens, Mangans und Nickels. Clausthaler Hefte zur Lagerstaettenkunde und Geochemie der Mineralischen Rohstoffe. 9; Pages 85-107. 1970. Notes: No. 9. Borntraeger. Berlin-Stuttgart, Federal Republic of Germany. Language: German; Summary Language: English. Abstract: Precambrian, regional metamorphism (570 degrees -630 degrees C, 5.5 kb pressure, 2910-3280 m.y.), local thermal metamorphism (680 degrees C, 3.5 kb pressure, 1600 m.y.), chemical composition (with analyses). Descriptors: Africa; alteration; Bong Range; chemical analysis; data; dates; gneisses; Liberia; metamorphic rocks; metamorphism; petrology; regional; schists; West Africa; Igneous and metamorphic petrology. ISSN: 0578-4697.

Stobernack, Just. 1968. Stratigraphie und Metamorphose des präkambrischen Grundgebirges der Bong Range in Liberia. Language: German. 185 pages: illustrations; 21 cm. Notes: Vita. Thesis (doctoral) - Technische Universitat Clausthal. Bibliography: p. 180-185. Descriptors: Metamorphism; Geology- Liberia; Bong Range. USGS Library; OCLC: 8722797

Stockwell, G. S. and Lugenbeel, J. W. 1868. The Republic of Liberia: Its Geography, Climate, Soil and Productions, With a History of its Early Settlement. New York, A.S. Barnes & Co. 1868. 298 pages. OCLC: 2720395.

Storette, Ronald F. 1971. The politics of integrated social investment. An American study of the Swedish LAMCO project in Liberia. Stockholm, Läromedelsförlagen. 108 pages, 23 cm. Series: Scandinavian university books. Descriptor: Technical assistance, Swedish -- Liberia. Notes: Includes bibliographical references. With a foreword by Folke Schmidt. LCCN: 70-593611; OCLC: 198634.

Storti, C. 1981. Forestry Case Studies. Peace Corps, Washington, DC. Information Collection and Exchange. Performer: TransCentury Corp., Washington, DC. 104 pages. Notes: Prepared in cooperation with TransCentury Corp., Washington, DC. Descriptors: History; Morocco; Nepal; Philippines; Chile; Guatemala; Chad; Liberia; Niger. *forests; developing country application; Peace Corps. Abstract: The report includes historical evaluations of eight forestry programs in Morocco, Nepal, the Philippines, Chile, Guatemala, Chad, Liberia and Niger. It also includes programming suggestions based on the studies and a list of forestry projects in other Peace Corps countries. NTIS Number: PB85239119XSP.

Stott-Cooper, H. 1967. "Some notes on the damming operations carried out by the Liberian Swiss Mining Corporation on the Lofa River." Bulletin of the Geological, Mining and Metallurgical Society of Liberia. 2; Pages 9-16. 1967. Geological, Mining and Metallurgical Society. Monrovia, Liberia. Descriptors: Africa; damming operations; diamonds; economic geology; Liberia; Lofa River; placers; production; river bed; West Africa; Economic geology, general. Illustrations, including sketch maps. ISSN: 0367-4819.

Stott-Cooper, H. 1966. "Some Notes on the Damming Operation Carried Out by the Liberian Swiss Mining Corporation on the Lofa River." Paper presented at the March monthly meeting of the Liberian Geological Society in 1966. Published in Annual Publication, p. 9-35. Also published in: Geological, Mining & Metallurgical Society of Liberia. Bulletin. Volume 2, pages 9-16. AMRS, Inc. See: http://www.africaminerals.com/

van Straaten, Peter. 2002. Rocks for Crops: Agrominerals of sub-Saharan Africa. International Centre for Research in Agroforestry (ICRAF), Nairobi, Kenya. 338pp. Printed by Fidelity National Information Solutions Canada, Scarborough, ON, M1B 3C3, Canada. Descriptors: sustainable agriculture; soil fertility management; mineral resources; sub-Saharan Africa. Abstract: There are two small known phosphate deposits in Liberia. The deposit at the Bomi iron ore deposit, 60 km north of Monrovia, is composed of grey to cream coloured phosphatic rocks superficially resembling calcareous tufa. Secondary Fe-phosphate, mainly phosphosiderite and strengite, fill cavities and form cements between iron ore fragments. A sample from the leucophosphite $(KFe_2[PO_4]_2OH(H_2O)_2)$ at Bomi contained 33.46% P_2O_5 and 36.85% Fe_2O_3, as well as 7.86% K_2O. Axelrod *et al.* (1952) interpreted this occurrence as a reaction product of bat excreta with iron oxides. There is an abundance of bat guano in the caves. No figures exist on volume and grade of the Bomi phosphate resources. The Bambuta deposit is located 70 km north-northeast of Monrovia, 25 km east of Bomi Hill iron mine (6°56' N; 10° 33' W). Results of a diamond drilling program and mapping have shown a minimum reserve of 1 million tons of phosphate rock at 32% P_2O_5, or 1.5

million tons of ore grading 28% P2O5 (Rosenblum and Srivastava 1979). The phosphorus bearing minerals are mainly members of the variscite-strengite series (secondary Al-Fe phosphates). Like at Bomi, the phosphates are associated with an iron ore deposit. The genesis of this deposit remains unclear although Rosenblum and Srivastava (1979) discuss the possibilities of a metasedimentary-metasomatic origin or, alternatively, origin as a result of phosphate precipitation from guano-derived solutions. ISBN 0-88955-512-5. See:
http://www.uoguelph.ca/~geology/rocks_for_crops/33liberia.PDF

Subah, P. 1981. "Iron ore in Liberia; past production and future prospects." Mining Magazine (London). 145; 3, Pages 204-208. 1981. Mining Journal Ltd.. London, United Kingdom. Abstract: A brief history of the iron ore mineral industry in Liberia. Description of current exploitations including production, production capacities and exports. Performance of companies. Outlook: the world economic situation is not favorable for expansion of the industry; the Nimba Mifergui project could prolong LAMCO's activities, whereas Liberia Mining Co. (LMC) will have to interrupt the feasibility study of Bea Mountains' deposits. In 1980 Liberia produced 18.98 million tons of iron ore and exported 18.80 million tons, 70 percent to the EEC countries, 19 percent to other European and Asian countries and 11 percent to the United States. Descriptors: Africa; companies; economic geology; economics; export; history; inventory; iron ores; LAMCO; Liberia; metal ores; mineral resources; production; programs; West Africa. ISSN: 0308-6631.

"Summary of Synoptic Meteorological Observations (SSMO). West African and Selected Island Coastal Marine Areas. Volume 2. Area 9, Conakry. Area 10, Monrovia. Area 11, Ivory Coast. Area 12, Accra. Area 13, Gulf of Guinea East, Area 14, Luanda NW. Area 15, Lobito." 1976. Naval Weather Service Detachment Asheville N C, United States. November 1976. Pages: 564. Descriptors: meteorological data; synoptic meteorology; West Africa; coastal regions; weather; tables (data); marine climatology; sea states; periodic variations; Angola; Ivory Coast. Identifiers: Guinea, Liberia, Gulf of Guinea. Abstract: This report presents marine climatological data for specific coastal areas in 21 different tables including weather occurrence, wind direct and speed, cloud amount, ceiling height, visibility, precipitation, dry bulb, relative humidity, air-sea temperature difference, sea height and period, sea surface temperature and sea level pressure. Limitation Code: Approved for public release. DTIC Number: ADA031779. Supplementary Note: See also Volume 1, AD-A031 778.

Summerfelt, R. L. 1967. "Preliminary operational report for the detailed aeromagnetic and radiometric survey of the Republic of Liberia." Bulletin of the Geological, Mining and Metallurgical Society of Liberia. 2; Pages 86-93. 1967. Geological, Mining and Metallurgical Society. Monrovia, Liberia. Abstract: Iron ore deposits, exploration. Descriptors: Africa; airborne; economic geology; exploration; geophysical methods; geophysical surveys; iron; iron exploration; Liberia; magnetic methods; metals; radioactivity methods; regional; regional aeromagnetic; surveys; West Africa; economic geology of ore deposits. Illustrations. ISSN: 0367-4819.

Szczesniak, Philip A. 2000. "The Mineral Industries of Côte D'Ivoire, Guinea, Liberia, and Sierra Leone." Abstract: Mineral production in Liberia consisted mainly of artisanal recovery of diamond and gold. The Government encouraged investment in natural resource development, although formal economic activity has been slow to recover since the end of the civil war in 1997. International exploration increased in 2000 as companies came to Liberia to examine what has been described as one of the last areas of virtually unexplored rocks in the world. Eastern Liberia is made up of rocks of Birimian age with significant potential for gold. Western Liberia is made up of rocks of Archean age that contain diamond, gold, iron ore, nickel, manganese, palladium, platinum, and uranium. See: http://minerals.usgs.gov/minerals/pubs/country/2000/ivgvlislmyb00.pdf.

Tarpeh, James Teah. 1978, 1979. The Liberian-LAMCO Joint Venture partnerships: the future of less developed country and multinational corporation collaboration as a national strategy for host country development. Thesis (Ph. D.)--University of Pittsburgh, 1978. 641 leaves : ill., maps. Descriptors: investments, foreign – Liberia; iron industry and trade – Liberia; economic development projects -- Liberia. Named Corporation: LAMCO. Notes: Includes bibliographical references (leaves 609-620). OCLC: 24858547.

Tassell, Arthur. 2001. "Mano River pioneers a new mining frontier in Liberia." African Mining. 6; 2, Pages 18-25. 2001. Descriptors: Africa; drilling; gold ores; Liberia; Mano River; metal ores; mineral exploration; mining; mining geology; production; West Africa; economic geology, geology of ore deposits. MAP COORDINATES: LAT: N040000; N080000; LONG: W0070000; W0120000. Illustrations, including sketch map.

Taylor, Wayne Chatfield. 1956. The Firestone Operations in Liberia. Washington: National Planning Association. 115 pages. Cover title: United States Business Performance abroad: The Case Study of the Firestone Operations in Liberia. Abstract: The fifth in an NPA series of United States business performance abroad, this is a study of one of the first American experiments in investment and large-scale plantation operations in tropical Africa. The author devotes a beginning third of his text to "Liberia- Past and Present," telling briefly significant facts of Liberia geography, history, economic, social and political development. Then he recites the steps leading to the Firestone concession of 1926, with its three agreements of harbor, planting and loan, and explains the development, operation and organization for the plantations, the care of Liberian workers, the arrangements for American and European staff, and subsidiary activities of Firestone. The last section evaluates Firestone's impact on the country and the future outlook, both for the country and the progress of Liberia. OCLC: 360147. Reprinted 1976: ISBN: 0405092873; OCLC: 2238145.

Taylor, A.; Boukambou, G.; Dahniya, M.; Ouayogode, B. V. and Ayling, R. I. 1996. "Strengthening National Agricultural Research Systems in the Humid and Sub-Humid Zones of West and Central Africa." A Framework for Action. World Bank Technical Paper No. 318. Performer: World Bank Group, Washington, DC. April 1996. 120p. NTIS Number: WB0630XSP. Order this product from NTIS by: phone at 1-800-553-NTIS (U.S. customers); (703)605-6000 (other countries); fax at (703)321-8547; and email at orders ntis.fedworld.gov. NTIS is located at 5285 Port Royal Road, Springfield, VA, 22161, USA. Descriptors: tropical-agriculture; agricultural-innovations; Benin; Cameroon; Central African-Republic; Congo; Cote d'Ivoire; Equatorial-Guinea; Gabon; Ghana; Guinea; Guinea Bissau; Liberia; Nigeria; Sao Tome and Principe; Sierra Leone; Togo; Zaire; *agricultural research; agriculture; Sub Saharan Africa. Abstract: This paper details a framework for action (FFA) to strengthen agricultural research in 17 countries of the humid and sub-humid zones of West and Central Africa. The framework includes a strategy, guidelines, and a process that enable agricultural research to generate the technology required to increase agricultural productivity while conserving the resource base. It is based on the building of coalitions among national, regional, and international stakeholders in technology generation and transfer. ISBN: 0821335669.

Terpstra, H. 1937. Geologische notities over Liberia; met petrografische beschrijvingen van Ch. E. A. Harloff. Translated "Brief account of the geology of Liberia, and short descriptions of rocks collected." De Ingenieur in Nederl.-Indiee, Jg. 4, IV. 7; Pages 136-139. 1937. Language: Dutch. Abstract: Brief account of the geology of Liberia, and short descriptions of rocks collected. Descriptors: Africa; igneous rocks; Liberia; West Africa; igneous and metamorphic petrology. Illustrations: 2 figs.

Terpstra, H. 1939. "Diamonds in Gold Coast Colony." Mining Magazine (London). 60; 4, Pages 219. 1939. Mining Journal Ltd.. London, United Kingdom. 1939. Abstract: A short note on diamonds in Liberia apropos of a previous article in the same journal on Gold Coast Colony diamonds. Descriptors: Africa; diamonds; gems; Liberia; West Africa; economic geology of nonmetal deposits. ISSN: 0308-6631.

Terrier, A. 1912. "Les Frontières de la Guinée Française de la Sierra Leone et du Liberia." Translated title: "The Borders of Guinea, Sierra Leone and Liberia." L'Afrique Française. Volume 22, pages 335-340.

Testa, Stephen Michael. 1983. "Geochemistry of Mesozoic tholeiites from Liberia, Guyana and Surinam." In: Geological Society of America, Southeastern Section, 32nd annual meeting. Abstracts with Programs - Geological Society of America. 15; 2, 1983. Geological Society of America (GSA). Boulder, CO, United States. Pages: 91. Conference: 32nd annual meeting, Southeastern Section, Geological Society of America; with the Southeast Section of the National Association of Geology Teachers and the Southeastern Section of the Paleontological Society. Tallahassee, FL, United States. March 16-18, 1983. Descriptors: Africa; age; alkaline earth metals; Atlantic Ocean; basalts; Central Atlantic Ocean; diabase; dikes; geochemistry; Guyana; Guyana Shield; igneous rocks; intrusions; iron; Liberia; magnesium; manganese; Mesozoic; metals; northwestern Liberia; Paleozoic; Pan African Orogeny; Paynesville Sandstone; Permian; petrography; petrology; phosphorus; plutonic rocks; Precambrian; Proterozoic; silica; South America; Surinam; tholeiite; titanium; Triassic; upper Precambrian; volcanic rocks; West Africa. ISSN: 0016-7592.

Testa, Stephen Michael. 1978. Geochemistry of Mesozoic dolerites from Liberia, Africa and Spitsbergen. Master's thesis, California State University, Northridge, CA, United States. Pages: 112. Descriptors: Africa; Arctic region; chemical composition; diabase; dikes; geochemistry; igneous rocks; intrusions; Liberia; major elements; Mesozoic; minor elements; petrology; plutonic rocks; Spitsbergen; Svalbard; West Africa; Geochemistry of rocks, soils, and sediments; Igneous and metamorphic petrology.

Testa, Stephen M. and Weigand, Peter W. 1980. "Petrochemistry of Mesozoic tholeiite intrusions from Liberia, West Africa, and comparison with other circum-Atlantic tholeiite provinces." In: The Geological Society of America, 93rd annual meeting. Abstracts with Programs - Geological Society of America. 12; 7, Pages 534. 1980. Geological Society of America. Boulder, CO, United States. Conference: 93rd annual meeting of the Geological Society of America. Atlanta, GA, United States. November 17-20, 1980. Descriptors: Africa; basalts; chemical composition; diabase;

differentiation; dikes; eastern North America; igneous rocks; intrusions; Liberia; Mesozoic; mineral assemblages; Morocco; North Africa; North America; Paleozoic; Pan African Orogeny; Paynesville Sandstone; petrology; plutonic rocks; Precambrian; Proterozoic; tholeiite; upper Precambrian; volcanic rocks; West Africa; igneous and metamorphic petrology. ISSN: 0016-7592.

Thayer, Thomas P. 1953. "The iron deposits of western Liberia." Report of the ... Session - International Geological Congress. Pages 49. 1953. Notes: 19th, Algeria, C. R. sec. 10, f. 10 International Geological Congress. Descriptors: Africa; iron; Liberia; metals; west; West Africa. ISSN: 1023-3210.

Thayer, Thomas P. 1952. "Iron ore deposits of Liberia." Economic Geology and the Bulletin of the Society of Economic Geologists. 47; 7, Pages 777-778. 1952. Economic Geology Publishing Company. Lancaster, PA, United States. Descriptors: Africa; iron; Liberia; Liberia, Bomi hills Bambuta area, phosphate mineralization; metals; mineral deposits, genesis; West Africa. Geol. Soc. Am., B. v. 63, no. 12, pt. 2, p. 1303. ISSN: 0361-0128.

Thayer, Thomas P.; Lill, G.G. and Coonrad, W. L. 1974. Mineral exploration in western Liberia; 1949-1950. U. S. Geol. Survey; U. S. Department of Commerce. Special Papers - Liberia, Geological Survey. 4, 1974. Monrovia, Liberia. Pages: 51; illustrations, maps; 27 cm. Series: Liberian Geological Survey Special paper; no. 4. Notes: Bibliography: p. 50-51. Descriptors: 1949-1950; Africa; bauxite; chromite ores; copper ores; corundum; diamonds; economic geology; folds; gold ores; granites; history; igneous rocks; iron ores; itabirite; lead ores; Liberia; manganes; metal ores; metamorphic rocks; mineral exploration; mineral resources; ore deposits; oxides; placers; plutonic rocks; Precambrian; schists; West Africa; zinc ores; Economic geology, general. References: 20; tables, sketch maps. ISSN: 0375-6831; OCLC: 4207663.

Theurkauf, E. 1972. "Stufenweise Durchfuehrung von Lagerstaettenuntersuchungen unter dem Gesichtspunkt der Wirtschaftlichkeit am Beispiel der Eisenerzlagerstaette, Mano River, Liberia, Westafrika." Translated title: "The stages of investigations of ore deposits from the economic point of view using the iron ore deposits at Mano River, Liberia, western Africa, as an example." Geologie en Mijnbouw. 51; 4, Pages 473-486. 1972. De Bussy Ellerman Harmanuscript Amsterdam, Netherlands. Language: German. Descriptors: Africa; composition; economic geology; iron ores; Liberia; Mano River; metal ores; methods; ore deposits; Precambrian; production; structure; West Africa; Economic geology of ore deposits; Illustrations, including geological sketch maps. ISSN: 0016-7746.

Thienhaus, Rolf; Eichler, Jurgen and Quade, Horst, eds. 1970. Montangeologische Untersuchungen an Lagerstatten des Eisens, Mangans und Nickels Translated title: Mining geological investigations of iron, manganese and nickel. Zur Erinnerung an Prof. Dr. Rolf Thienhaus. Hrsg. von J(urgen) Eichler u. H. Quade. Mit 18 Photos, 71 Abb. u. 50 Tab. im Text u. auf 6 Beil. Berlin, Stuttgart, Borntraeger, 1970. Description: 244 pages with illustrations, maps. 25 cm. Contents: Die geologische Position der prakambrischen Quarzbandererze (Itabirite) und die Problematik ihrer Genese, von J. Eichler.--Der Bildungsraum und die genetische Problematik der vulkano-sedimentaren Eisenerze, von H. Quade.--Metamorphose und Genese zentralindischer Manganerze, von H. D. Fuchs.--Metamorphose der Glimmerschiefer und Gneise in der westlichen Bong Range, Liberia, von J. Stobernack.--Montangeologische Grundiagen der betriebsgeologischen

Abbausteuerung auf der Eisenerzlagerstatte Bong Rang in Liberia/Westafrika, von J. Lersch.--Lateritische Nickelerze, von H. Gruss. Zur Entstehung und Charakteristik mesozoischer marinsedimentarer Eisenerze im ostlichen Niedersachsen, von H. Kolbe.--Feinstratigraphische und petrographische Untersuchungen im erzfuhrenden Korallenoolith (Unterer Malm) des westlichen Wesergebirges (Nordwestdeutschland), von K.--P. Freitag.--Lithostratigraphie und Biochrolologie im Grenzbereich Oxford/Kimmerdge (Malm) des Wesergebirges und Suntels (NW-Deutschland), von W. Schiebel. Notes: Includes bibliographies. Descriptors: ore deposits; Geology. Series: Clausthaler Hefte zur Lagerstattenkunde und Geochemie der mineralischen Rohstoffe, Heft 9. ISBN: 3443120091.

Thienhaus, R. and Stobernack, J. 1967. "Investigations on facies, metamorphism, and tectonics of the Precambrian basement of the Bong Range, Liberia." Bulletin of the Geological, Mining and Metallurgical Society of Liberia. 2; Pages 48-58. 1967. Geological, Mining and Metallurgical Society. Monrovia, Liberia. Descriptors: Africa; anatexis; Bong Range; Liberia; metamorphic rocks; metamorphism; metasediments; Metavolcanics; mineral composition; mineral facies; petrology; Precambrian; Precambrian basement; tectonics; West Africa; Igneous and metamorphic petrology. Illustrations, including geological sketch maps. ISSN: 0367-4819.

Thilmans, G.; Regnaud, M.; Hebrard, L.; Guitat, R. and Descamps, C. 1974. "Fichier des ages absolus du Quaternaire d'Afrique au Nord de l'Equateur; 9 (super e) Serie." Translated title: "Index cards of absolute ages from the Quaternary of Africa north of the Equator; 9th series." Bulletin de Liaison - Association Senegalaise pour l'Etude du Quaternaire de l'Ouest Africain. 40; Pages 5-131. 1974. Notes: No. 40. Association Senegalaise pour l'Etude du Quaternaire Africain. Dakar, Senegal. Language: French. Descriptors: absolute age; Africa; Algeria; Atlantic Ocean; Atlantic Ocean Islands; C-14; Cameroon; Canary Islands; carbon; Cenozoic; Central Africa; Central African Republic; Chad; Congo; data; dates; East Africa; Egypt; Ethiopia; geochronology; Ghana; Guinea; isotopes; Ivory Coast; Kenya; Liberia; Libya; Mali; Mauritania; Morocco; Niger; Nigeria; north; North Africa; organic compounds; organic materials; Quaternary; radioactive isotopes; sediments; Senegal; Sudan; Tunisia; Uganda; West Africa; wood; geochronology. ISSN: 0044-9725.

Thomson, J.; Kanaan, R.; Baker, M.; Clausen, R.; N'Goma, M. and Roule, T. 2004. "Conflict Timber: Dimensions of the Problem in Asia and Africa. Volume Three: African Cases." Performer: ARD, Inc., Burlington, VT, United States. Sponsor: Agency for International Development, Washington, DC. Bureau for Asia and Near East. 2004. one CD-ROM contains 164 page document. Descriptors: Africa; wood products; origins; objectives; task order implementation; analytical framework; commodities; markets; forest resources; Gabon; Congo; Guinea; Liberia; Sierra Leone; governance; comparisons; trends. Abstract: Conflict timber-conflict financed or sustained through the harvest and sale of timber (Type 1), or conflict emerging as a result of competition over timber or other forest resources (Type 2)-poses serious problems in many countries in Asia and Africa. While forest resources, particularly timber, are far from the only commodities that spark or finance conflicts, they have certainly played a considerable role in sustaining many conflicts in these regions. Conflict timber as defined in this study occurs in several settings in contemporary Africa. The Conflict Timber Study Team targeted two types of conflict timber incidents: Conflicts fueled by forest resources, and conflicts emerging because of competition over forest resources. This volume contains an in-depth study of conflict timber in the

Democratic Republic of Congo (DRC) as well as shorter, comparative desk studies of conflict timber in several other countries-among them Liberia. The latter country is clearly the 'poster child' of African conflict timber. The country's president, Charles Taylor, has authorized logging concessionaires to harvest timber which he has then exchanged with external partners to obtain the means of war light arms, helicopter parts, etc. These transactions have involved both timber sales, typically of raw logs, as well as logs bartered directly in exchange for arms. The People's Republic of China (PRC) has been a prominent Liberian barter partner in the recent Overview of Conflict Timber in Africa past. Taylor has used the timber-based military resources he has acquired to maintain his power at home and to pursue wars against Liberia's immediate neighbors (Sierra Leone, Guinea and, most recently, the Ivory Coast). Notes: This document is color dependent and/or in landscape layout. It is currently available on CD-ROM and paper only. CD-ROM contains a 164 page document. See also PB2005-102558 and PB2005-101559. Sponsored by the Agency for International Development, Washington, DC. Bureau for Asia and Near East. NTIS: PB2005102560.

Thompson, K. and Crawshaw, J. 1998. "Water in Liberia - how war affects policy formulation and implementation." Waterlines 16, no. 3 (1998) p. 27-29. Descriptors: water resources; water policy; administrative agencies; water management; developing countries. Abstract: Liberia suffered 7 years of civil war – 150,000 people died and the infrastructure was practically destroyed. But now there is a real window of opportunity for co-operation between poor, but experienced ministries, agencies, and donors. Subjects: Water supply and sanitation. ISSN: 0262-8104.

Thorman, Charles H. 1988. "Tectonic setting of Liberia." In: The West African connection; evolution of the central Atlantic Ocean and its continental margins. Sougy, Jean (editor); Rodgers, John (editor). Journal of African Earth Sciences. 7; 2, Pages 515. 1988. Pergamon. London-New York, International. Conference: The West African connection; evolution of the central Atlantic Ocean and its continental margins. Giens, France. January 17-22, 1984. Descriptors: Africa; facies; folds; granulite facies; greenschist facies; Liberia; metamorphism; regional metamorphism; structural geology; synclines; tectonics; Todi shear zone; uplifts; West Africa. ISSN: 0731-7247.

Thorman, Charles H. 1977. "Geologic map of the Monrovia Quadrangle, Liberia." Miscellaneous Investigations Series Map. Report number: I-0775-D. January 1, 1977. U. S. Geological Survey, Reston, VA, United States. Map Type: geologic map. Bibliographic level: Monograph color map; 44 x 66 cm. on sheet 74 x 129 cm. folded in envelope 30 x 24 cm. Map Scale 1:250,000 (1 inch = about 4 miles). Sheet 29 by 51 inches. Descriptors: Africa; areal geology; geologic maps; Liberia; maps; Monrovia Quadrangle; USGS; West Africa. Notes: "Hotines rectified skew orthomorphic projection and rectified coordinates." "Prepared by the U.S. Geological Survey and the Liberian Geological Survey under the joint sponsorship of the Government of Liberia and the Agency for International Development, U.S. Department of State." Includes bibliography. ISSN: 0160-0753; OCLC: 3594270.

Thorman, Charles H. 1976. "Implication of klippen and a new sedimentary unit at Gibi Mountain, Liberia, West Africa, in the problem of the Pan-African--Liberian age province boundary." Geological Society of America Bulletin. 87; 6, Pages 851-856. 1976. Geological Society of America (GSA). Boulder, CO, United States. Descriptors: Africa; allochthons; fault zones; faults; folds; Gibi Mountain; klippen; Liberia; Liberian; lithostratigraphy; lower Paleozoic; orogeny;

Paleozoic; Pan African; Precambrian; sedimentary rocks; stratigraphy; structural complexes; structural geology; tectonics; thrust; West Africa; Structural geology. ISSN: 0016-7606.

Thorman, Charles H. 1974. "Geologic map of the Monrovia Quadrangle, Liberia." Open-File Report. Date: January 1, 1974. Report number: OF 74-0305. Pages (monograph): 18 pages (1 sheet). U. S. Geological Survey, Reston, VA, United States. Annotation: 1 sheet, scale 1:250,000 (1 inch = about 4 miles). Descriptors: Africa; areal geology; geologic maps; Liberia; maps; Monrovia Quadrangle; USGS; West Africa. Notes: Map in pocket. Prepared under the auspices of the Govt. of Liberia and the Agency for International Development, U.S. Dept. of State. Transmittal sheet dated December 13, 1974. Bibliography: leaves 17-18. Subjects: Geology- Liberia- Monrovia Quadrangle. Mines and mineral resources- Liberia- Monrovia Quadrangle. Monrovia Quadrangle (Liberia). Notes: "Prepared under the auspices of the Government of Liberia and the Agency for International Development, U.S. Department of State"--Cover. "Part of a program undertaken cooperatively by the Liberian Geological Survey (LGS) and the U.S. Geological Survey (USGS)"-- P. [1]. "Project report, Liberian investigations, (IR) LI-66B"--Cover. One folded map in pocket. Includes bibliographical references (p. 17-18). Other Titles: Monrovia Quadrangle, Liberia. ISSN: 0196-1497.

Thorman, Charles H. 1973. A field trip in the vicinity of Monrovia, Liberia. [Washington]: U.S. Dept. of the Interior, Geological Survey. Description: 14 leaves: maps (2 folded in pocket); 30 cm. Descriptors: Geology- Monrovia: Liberia- Guidebooks. Notes: "Project report Liberian investigations (IR) LI-81." "Prepared under the auspices of the Agency for International Development, U.S. Department of State, and the Government of Liberia." Bibliography: leaf 14. LCCN: 78-309821; OCLC: 4490865.

Thorman, Charles H. 1972. "The boundary between the Pan-African and the Liberian Age provinces, Liberia, West Africa." Abstracts with Programs - Geological Society of America. 4; 7, Pages 690. 1972. Geological Society of America (GSA). Boulder, CO, United States. Descriptors: Africa; effects; faults; Liberia; Liberian; Paleozoic; Pan-African; shear zones; structural geology; tectonics; West Africa. ISSN: 0016-7592.

Timmerman, Kenneth R. 2001. "Severing Liberia's sinews of war: the U.N. is preparing new sanctions against Liberia's strongman Charles Taylor, who seized that African country's profitable international shipping registry and is brokering `conflict diamonds' from guerrillas. (World: Sanctions) (Brief Article)." Insight on the News. November 19, 2001: pages 26 et seq. Abstract: In May 2000, the British government sponsored a U.N. Security Council (UNSC) resolution to impose a worldwide ban on the import of conflict diamonds from Liberia, Sierra Leone and other African states, but the Clinton administration- guided by Jackson and Payne- opposed it. A new resolution (UNSC Resolution 1343) passed with strong support from the Bush administration on March 7 and became effective two months later. In addition to the worldwide ban on conflict diamonds, it included a ban on travel by senior Liberian government officials and their spouses, as well as any other individuals "providing financial and military support to armed rebel groups in countries neighboring Liberia." But Taylor and his arms and diamond brokers continue their deadly trade. New on the radar screen are Liberia's growing timber exports, which Taylor is exploiting as a source of revenue and to provide cover for arms imports. In one particular deal in May, he attempted to purchase from Libya several thousand AK-47 assault rifles and ammunition and RPG-

7 rocket-propelled grenades. In some cases, "these arms arrive on some of the same ships that subsequently export logs," according to a recent report by Global Witness, an NGO based in London (www.oneworld .org/globalwitness/liberia/liberiareport cover.htm). In 2000, Liberia exported timber worth $187 million on the world market. And yet the Liberian government declared timber revenues for the same period worth a scant $6.7 million. The U.N. experts believe Taylor used the rest of the money to finance his personal lifestyle and war. U.N. Secretary-General Kofi Annan nonetheless appears to be taking Jackson's view that more sanctions would just create more suffering in Liberia. But Liberian opposition leader Gayah Fahnbulleh scoffs at them both. "With or without timber, people are suffering. They are destitute, impoverished. So sanctions on timber will make no difference to the plight of ordinary Liberians. But they will mean the end of Charles Taylor, because all the money goes to Taylor and his inner circle."

Toft, Paul B.; Hills, Doris V. and Haggerty, Stephen E. 1989. "Crustal evolution and the granulite to eclogite transition in xenoliths from kimberlites in the West African Craton." In: Growth of the continental crust. Ashwal, Lewis D. (editor). Tectonophysics. 161; 3-4, Pages 213-231. Elsevier. Amsterdam, Netherlands. Conference: Workshop on The growth of continental crust. Oxford, United Kingdom. July 13-16, 1987. Descriptors: Africa; chemical composition; continental crust; crust; eclogite; geochemistry; granulites; igneous rocks; inclusions; kimberlite; Koidu; Liberia; major elements; mantle; metamorphic rocks; mineral composition; Mohorovicic discontinuity; plutonic rocks; Sample Creek; Sierra Leone; ultramafics; upper mantle; West Africa; West African Shield; xenoliths. Abstract: A petrographic, mineral and bulk chemical study of a xenolith suite of granulites and eclogites from Sample Creek, Liberia and Koidu, Sierra Leone, has been undertaken with a view to determining the nature of the crust-upper mantle interface. A broad range of xenolith compositions is present (from high-MgO eclogites to garnet-anorthosites), and a systematic AFM trend is established, consistent with mafic and ultramafic melt fractionation at moderate pressures (10-20kbar). A trend is established for the entire xenolith suite among bulk chemistry, seismic P-wave velocity, and a crust/mantle (C/M) bulk chemical ratio defined as $Na_2O + K_2O + SiO_2/FeO + MgO$ mole %. Three populations are present: a granulite crustal group; a granulite and eclogite transitional group; and an exclusively upper mantle eclogitic group. From these data, and coupled with garnet-clinopyroxene mineral thermometry and accessory phases (eg, diamond, graphite, coesite, kyanite) or the presence of plagioclase, a xenolith geotherm is established based on stratigraphic sequencing and phase transition boundaries. Diamond and coesite-bearing eclogites conform to the $40mW/m2$ standard cratonic low heat flow geotherm, whereas the plagioclase granulites at lower pressures correspond to an average rift geotherm of $90mW/m2$. The latter is ascribed to igneous underplating onto the lower crust or to thermal perturbations from an earlier tectonic event. Graphite and kyanite ecologites and the transitional group (in SG, Vp and C/M ratio) of eclogites and granulites fall between the 40 and $90mW/m2$ reference geothermanuscript. References: 89; illustrations, including 5 tables, sketch map. ISSN: 0040-1951.

Tooms, J. S. 1987. "Exploration for gold in the humid tropics." Journal of Geochemical Exploration. 1987. Descriptors: gold; rivers; sediments; geochemical surveys; surveys; tropical environments; South America; Africa, West. Abstract: Exploration for hard-rock Au deposits has been undertaken by the United Nations Revolving Fund for Natural Resources Exploration. (UNRFNRE) in a considerable number of countries. This paper summarizes the experience gained in areas of tropical rain forest in northern South America and in West Africa and is illustrated by results from Suriname and Liberia in particular. Comparative studies have been undertaken of prospecting by panning stream sediment and by analyzing the stream sediment fines (-80 mesh fraction). Gold mineralized zones crossed by a number of streams are generally outlined by both methods. However, results at individual sites, not uncommonly, are highly divergent. ISSN: 0375-6742.

Topographic Section, General Staff. Various dates. "Liberia." Notes: Parts of Sheets 59 and 71. "2083." Scale: 1:1,000,000. 1905 (409B). 1914. Library of Congress, Geography & Map Division.

Toweh, Solomon Hartley. 1989. Prospects for Liberian iron ores considering shifting patterns of trade in the world iron ore industry. Dissertation: Thesis (Ph. D.). Mining and Geological Engineering, University of Arizona, 1989. 356 leaves: ill., maps; 28 cm. Subjects: Iron ores-Economic aspects- Liberia. Economic forecasting- Liberia. International trade. Liberia- Economic conditions. Descriptors: Africa; demand; economic geology; economics; history; industry; iron ores; Liberia; markets; metal ores; quantitative analysis; supply; West Africa. Notes: Includes bibliographical references. Reproduction: Microfiche. Ann Arbor, Mich.: University Microfilms International, 1989. 4 microfiches: ill., maps. OCLC: 22213427.

Transportation map of Liberia. 1976. Liberian Cartographic Service. [Monrovia]: The Service. 1 map: photocopy; 71 x 57 cm. Subjects: Transportation- Liberia- Maps. Scale 1:1,000,000. Responsibility: prepared by Liberian Cartographic Service, Ministry of Lands & Mines, Apr. 1976. Notes: Blueline print. Shows roads, railroads, 1st order administrative divisions, rivers, iron ore mines, and rubber plantations. Includes distance chart. LCCN: 85-690694; OCLC: 12836252; 23269033.

Tricart, Jean. 1962. "Observations de geomorphologie littorale a Mamba Point (Monrovia, Liberia); Geomorphologische Beobachtungen am Kap Mamba Point." Erdkunde. 16; 1, Pages 49-57. 1962. Ferd. Duemmlers Verlag. Bonn, Federal Republic of Germany. Language: French; Summary Language: German. Abstract: Coastal erosion forms at Mamba Point, Liberia, are well developed at distinct elevations in relatively resistant dolerite. Salt spray retards vegetation at the upper edge; salt weathering, thermal scaling, and disintegration due to lower plant life proceeds above the water line, while solution channels are corroded at lower levels. Mamba Point is a transition type of tropical coast with relatively moist climatic conditions. Coastal terraces are cut nearer high-tide line than in temperate regions. Descriptors: Africa; erosion; Liberia; Mamba Point; shore features; shorelines; West Africa. ISSN: 0014-0015.

Troft, P. B. and Haggerty, Stephen E. 1985. "Magnetic susceptibilities and densities of rocks and minerals in the interpretation of geophysical anomalies." In: American Geophysical Union; 1985 spring meeting. Eos, Transactions, American Geophysical Union. 66; 18, Pages 256. 1985. American Geophysical Union. Washington, DC, United States. Conference: American Geophysical Union; 1985 spring meeting. Baltimore, MD, United States. May 27-29, 1985. Descriptors: Africa; anomalies; geophysical methods; granulites; interpretation; Josephine Creek; Liberia; magnetic methods; magnetic properties; magnetic susceptibility; magnetite; metaigneous rocks; metamorphic rocks; metasomatic rocks; Oregon; oxides; paleomagnetism; serpentinite; specific gravity; sulfides; United States; West Africa. ISSN: 0096-3941.

Tulk, Alfred J. 1933. "District No. 3 Central Province: Map composed from observations by Surveyor B. Powell & from routes followed by district Comm. G. Dunbar, Alfred J. Tulk & Dr. C. W. Harley." Prepared by a. J. Tulk, Santa Mission. February 1933. Scale unknown. Notes: Original tracing obtained from Mr. A. J. Tulk, 2 West 45th Street, NYC, through NYLB. Liberia, Northwest section. Blue line print. Library of Congress, Geography & Map Division.

Tverdokhlebov, V. A. 1970. "Tektonicheskoye stroyeniye zapadnoy chasti Gvineysko-Liberiyskogo shchita." Translated title: "Tectonic structure of the western part of the Guinea-Liberian shield. In: Voprosy tektoniki dokembriya kontinentov." Trudy Instituta Geologii i Geofiziki (Novosibirsk). 129; Pages 190-197. 1970. Notes: No. 129. Nauka, Sibirskoye Otdeleniye Instituta Geologii i Geofiziki. Novosibirsk, USSR. Language: Russian. Descriptors: Africa; evolution; Guinea Liberian shield; Precambrian; stratigraphy; Structural units; tectonics; west; geological sketch map. ISSN: 0568-658X.

Tysdal, Russell G. 1988. General geology of Liberia. In: The West African connection; evolution of the central Atlantic Ocean and its continental margins. Sougy-Jean (editor); Rodgers-John (editor). Journal of African Earth Sciences. 7; 2, Pages 515. 1988. Pergamon. London-New York, International. 1988. Conference: The West African connection; evolution of the central Atlantic Ocean and its continental margins. Giens, France. January 17-22, 1984. Descriptors: Africa; diabase; dikes; faults; gravity anomalies; igneous rocks; intrusions; Liberia; metamorphic rocks; plutonic rocks; Precambrian; shear zones; stratigraphy; Todi shear zone; West Africa. ISSN: 0731-7247.

Tysdal, Russell G. 1978. "Geology of the Juazohn quadrangle, Liberia."Liberian Geological Survey. Washington: U.S. Govt. Print. Office. 39 pages: maps; 23 cm. Notes: Prepared in cooperation with the Liberian Geological Survey under the sponsorship of the Agency for International Development, U.S. Department of State. Bibliography: p. 38-39. Descriptors: Liberian Geology-Juazohn. LCCN: 77-608150; OCLC: 3202615.

Tysdal, Russell G. 1978. "Geology of the Buchanan Quadrangle, Liberia: prepared in cooperation with the Liberian Geological Survey under the sponsorship of the Agency for International Development, U.S. Dept. of State." Liberian Geological Survey. USGS Bulletin. Report number: B 1449. Pages 31, maps; 23 cm. U. S. Geological Survey, Reston, VA, United States. January 1, 1978. Annotation: Prepared in cooperation with Liberian Geol. Survey under sponsorship of U. S. Dep. State, Agency Int. Dev. Illustrations, including geological sketch map. Number of references: 29. Descriptors: Africa; areal geology; Buchanan Quadrangle; Liberia; USGS; West Africa. Notes: Bibliography: pages 30-31. LCCN: 77-608145; OCLC: 3166667.

Tysdal, Russell G. 1978. "Geology of the Juazohn Quadrangle, Liberia." USGS Bulletin. Report number: B 1448. Date: January 1, 1978. Pages (monograph): 39. U. S. Geological Survey, Reston, VA, United States. Notes: Explanatory text for U. S. Geol. Surv. Misc. Invest. Map I-779-A. Annotation: Explanatory text for U. S. Geol. Surv. Misc. Invest. Map I-779-A. Illustrations, including geological sketch map. Descriptors: Africa; areal geology; chemically precipitated rocks; diorite family; diorites; explanatory text; faults; geologic maps; gneisses; igneous rocks; iron formations; iron ores; Juazohn Quadrangle; Liberia; maps; metal ores; metamorphic rocks; ore deposits; petrology; plutonic rocks; Precambrian; sedimentary rocks; structure; USGS; West Africa. References: 23; illustrations, including geol. sketch map. ISSN: 8755-531X.

Tysdal, Russell G. 1977. "Geologic map of the Buchanan Quadrangle, Liberia." Liberian Geological Survey. Miscellaneous Investigations Series Map. Report number: I-0778-D. U. S. Geological Survey, Reston, VA, United States. January 1, 1977. Map Type: geologic map; color map; 44 x 55 cm. on sheet 74 x 107 cm. folded in envelope 30 x 24 cm. Scale 1:250,000 (1 inch = about 4 miles). Sheet 29 by 42 inches. Descriptors: Africa; areal geology; Buchanan Quadrangle; geologic maps; Liberia; maps; USGS; West Africa. Notes: "Hotines rectified skew orthomorphic projection and rectangular coordinates." "Prepared by the U.S. Geological Survey and the Liberian Geological Survey under the joint sponsorship of the Government of Liberia and the Agency for International Development, U.S. Department of State." Includes bibliography. ISSN: 0160-0753; OCLC: 3600587.

Tysdal, Russell G. 1977. "Geologic map of the Juazohn Quadrangle, Liberia. Folio of the Juazohn quadrangle." Liberian Geological Survey. Miscellaneous Investigations Series Map. Report number: I-0779-D. U. S. Geological Survey, Reston, VA, United States. January 1, 1977. Map Scale: 1:250,000; geologic map; color map; 44 x 74 cm. on sheet 73 x 119 cm. folded in envelope 30 x 24 cm. Scale 1:250,000 (1 inch = about 4 miles). Sheet 29 by 47 inches. Descriptors: Africa; areal geology; geologic maps; Juazohn Quadrangle; Liberia; maps; USGS; West Africa. Notes: "Hotines rectified skew orthomorphic projection and rectangular coordinates." "Prepared by the U.S. Geological Survey and the Liberian Geological Survey under the joint sponsorship of the Government of Liberia and the Agency for International Development, U.S. Department of State." Includes bibliography. ISSN: 0160-0753; OCLC: 3600865.

Tysdal, Russell G. 1974. "Geology of the Buchanan Quadrangle, Liberia." Open-File Report. Report number: OF 74-0308. January 1, 1974. Pages: 14 (1 sheet). U. S. Geological Survey, Reston, VA, United States. Annotation: 1 sheet, scale 1:250,000 (1 inch = about 4 miles). Descriptors: Africa; areal geology; Buchanan Quadrangle; Liberia; USGS; West Africa. Notes: Map in pocket. Prepared in cooperation with the Govt. of Liberia and the Agency for International Development, U.S. Dept. of State. Transmittal sheet dated November 7, 1974. Bibliography: leaves 13-14. ISSN: 0196-1497.

Tysdal, Russell G. 1974. "Geology of the Juazohn Quadrangle, Liberia." Open-File Report. Report number: OF 74-0309. Date: January 1, 1974. Pages: 14 (1 sheet). U. S. Geological Survey, Reston, VA, United States. Annotation: 14 leaves: folded map; 27 cm. 1 sheet, scale 1:250,000 (1 inch = about 4 miles). Descriptors: Africa; areal geology; Juazohn Quadrangle; Juzohn Quadrangle; Liberia; maps; USGS; West Africa. Notes: Map in pocket. Prepared under the auspices of the Govt. of Liberia and the Agency for International Development, U.S. Dept. of State. Transmittal sheet dated December 9, 1974. Bibliography: leaves 13-14. ISSN: 0196-1497.

Tysdal, Russell G. and Thorman, C.H. 1983. "Geologic map of Liberia." Miscellaneous Investigations Series Map. Report number: I- 1480. Date: January 1, 1983. Pages (monograph): (1 sheet). U. S. Geological Survey, Reston, VA, United States. 1 map: color; 55 x 55 cm., sheet 69 x 107 cm., folded in envelope 30 x 24 cm. Map Scale: 1:1,000,000. Annotation: Lat 4 degrees to 9 degrees, long 7 degrees to 12 degrees. Scale 1:1,000,000 (1 inch = about 16 miles). Sheet 27 by 42 inches. Hotines rectified skew orthomorphic proj. (W 12-W 7/N 9-N 4). Descriptors: Africa; areal geology; geologic maps; Liberia; maps; USGS; West Africa. Notes: Text, "Age Province Map", and index map on sheet. "Prepared by the U.S. Geological Survey and the Liberian Geological Survey under the joint sponsorship of the Government of Liberia and the Agency for International Development, U.S. Department of State." Includes bibliography. ISSN: 0160-0753; OCLC: 10553595.

Underwood, David C. 1982. West African Oil: Will It Make a Difference? US Naval Postgraduate School, Monterey CA. Master's thesis. Report Date: Dec 1982, pages: 145. Descriptors: *economic analysis; *political science; *commerce; *petroleum products; *West Africa; developing nations; exports; production; economic impact; capacity (Quantity); theses; hydrocarbons; resource management; international trade; oils; petroleum industry; natural gas; crude oil; reserves(Energy); petroleum geology. Abstract: This thesis analyzes the commercial development of West Africa's petroleum resources and examines the implications of sudden 'oil wealth' for the region's political and economic development. Section one outlines the evolution of the petroleum industry and surveys the hydrocarbon potential of twenty nations along the coast from Senegal to Angola, and inland from Mali to the Central African Republic. An evaluation of the physical and political constraints on the development of the region's petroleum resources and an aggregate analysis of the area's potential for new oil reserves and production capacity are also presented. Finally, by drawing from the experiences of established oil producers like Nigeria, the economic and political implications of the widespread development of petroleum resources in West Africa are projected at the domestic and regional levels. Specifically, will new found oil wealth help resolve existing problems, or will it magnify them? The study concludes that for most of these countries, oil wealth will be a mixed blessing. Limitation Code: Approved for public release. DTIC Number: ADA123826. See: http://www.dtic.mil

United Nations. Department of Public Information. Cartographic Section. 1995. "Liberia- Proposed Team Deployment Sites." New York: The Section. 1 map, color, 16 x 21 cm. Scale 1:3,300,000. Notes: October 1996. Map No. 3879, rev. 1. LCCN: 97-680054.

United Nations. Department of Public Information. Cartographic Section. Various dates. "UNOMIL Deployment as of..." New York: The Section. Color maps, scale 1:3,300,000, 16 x 21 cm. (January 1996, Map No. 3802, rev. 15. LCCN: 97-680052) (March 1996, Map No. 3802, rev. 16. LCCN: 97-680053) (May 1996, Map No. 3802, rev 17. LCCN: 97-680060) (August 1996. Map No. 3802, rev. 18. LCCN: 97-680046) (October 1996, Map No. 3802, rev. 19. LCCN: 97-680061) (September 1997, Map No. 3802, rev. 25. LCCN: 97-687615).

United Nations. Economic Commission for Africa. 1964. "Map of Telecommunications Facilities in Liberia." Addis Ababa, Ethiopia: United Nations Economic Commission for Africa. 1 map, 24 x 20 cm. Scale 1:3,000,000. Notes: "Cart-A-64-14." LCCN: 96-687128.

United Nations Economic and Social Council. 1963. "Geological Bibliography of Africa- Bibliographie Geologique de l'Afrique." Economic Commission to Africa, Standing Committee on Industry, Natural Resources and Transport. Second Session, Addis Ababa, December 3-13, 1963. Liberia is page 161 and page 51 of the Annex. Report Number: E/CN.14/INR/48 15 October 1963. USGS Library.

United Nations Humanitarian Information Centre. 2003. Liberia (various titles and editions). URL: http://www.humanitarianinfo.org/liberia/ Abstract: The HIC is a common service to the humanitarian community working in Liberia and the surrounding countries. The HIC is managed by the United Nations Office for the Coordination of Humanitarian Affairs (UN OCHA), and operates in co-ordination with a number of partners including the United Nations High Commissioner for Refugees (UNHCR), the UN Joint Logistics Cell (UNJLC), the Department for Peacekeeping Operations (DPKO), the UK Department For International Development (DFID), USAID's Office of Foreign Disaster Assistance (OFDA) and Aid Workers Network. The HIC produces prints and distributes a wide range of Cartographic and Thematic maps, as well as providing hard copies of Satellite Images relating to Liberia. These map products are freely available via the website and at any HIC office. Some of these sets of map have been produced in USA; the Liberia HIC office has produced others. All these maps mainly provide information about administrative boundaries, populated places, road and rail network, and hydrography. These maps provide basic information about the country of Liberia, showing features such as administrative boundaries, road and rail networks, and populated places. The site also includes Landsat Mosaic and Landsat Topographic images. Ikonos City maps are based on very high resolution (1m) satellite imagery. Maps are available for most of the major urban areas of Liberia.

"UN imposes arms embargo on Liberia." 2001. Arms Control Today 31, no. 3 (April 2001): p. 31. ISSN: 0196-125X. Abstract: On March 7, the United Nations Security Council passed a resolution imposing a new arms embargo on Liberia for supporting the Revolutionary United Front (RUF), a rebel group that has waged a guerrilla war in Sierra Leone. Under the resolution, Liberia must also stop trading rough diamonds from Sierra Leon that are not controlled through the certificate-or-origin regime currently in place. Descriptors: embargoes & blockades; arms control; military sales; guerrilla forces; diamonds; mining.

U.S. Army Engineer Waterways Experiment Station; United States Army; Office of the Chief of Engineers. 1952. Limited reconnaissance for pavement evaluation and soil type-airphoto ties [sic]: report no. 13, Roberts Field, Liberia. Vicksburg, Mississippi: The Station. Description: iii, 23, [8] pages, 6 leaves of plates (1 fold.): ill.; 27 cm. Series: Technical memorandum; no.3-343, Rpt. 13; Variation: Technical memorandum (U.S. Army Engineer Waterways Experiment Station); no.3-343, Rpt. 13. Descriptors: runways (Aeronautics) - Liberia; pavements- Liberia; soils- Liberia; Roberts Field, Liberia. Notes: "December 1952." prepared by Waterways Experiment Station; for Office of the Chief of Engineers, Airfields Branch, Engineering Division, Military Construction. OCLC: 8117029.

United States. Army Map Service. Map of West Africa. Its Series G504; Variation: United States Army Map Service; A.M.S.; G504. color maps; 47 x 17 cm. or smaller; Scale 1:250,000. Notes: Issued in quadrangles 1^0 of latitude by $1^0$30' of longitude. "Transverse Mercator projection." Relief shown variously by contours, form lines, and shading. Each sheet includes "Glossary," "Location diagram," and "Reliability diagram." Some sheets have through route maps of cities on versos. Accompanied by index map on sheet 27 x 50 cm. LCCN: 56-381; OCLC: 6589872.

United States. Central Intelligence Agency. 1990. Liberia. Washington, D.C: Central Intelligence Agency, 1990. 1 map: color; 20 x 17 cm. MAP: Scale [ca. 1: 3,000,000] (W 12-W 7/N 9-N 4).

Notes: "Base 801529 (A02839) 6-90." OCLC: 22152577; USGS Library: M(754) v1990c.

United States. Central Intelligence Agency. 1990. Liberia. Washington, D.C: Central Intelligence Agency, 1990. 1 map: color; 20 x 17 cm. MAP: Scale [ca. 1: 3,000,000] (W 12-W 7/N 9-N 4). Notes: Relief shown by shading. "Base 801530 (A02839) 6-90." OCLC: 22152607.

United States. Central Intelligence Agency. 1987. Liberia. Washington, D.C: Central Intelligence Agency. 1 map: color; 21 x 17 cm. Scale [ca. 1: 3,000,000] (W 12-W 7/N 9-N 4). Notes: "Base 800638 (A02839) 5-87." LCCN: 87-692418 /MAPS; OCLC: 16952315.

United States. Central Intelligence Agency. 1987. Liberia. Washington, D.C: Central Intelligence Agency. 1 map: color; 21 x 17 cm. Scale [ca. 1: 3,000,000] (W 12-W 7/N 9-N 4). Notes: "Base 800639 (A02839) 5-87." Relief shown by shading. LCCN: 87-692419; OCLC: 17481370.

United States. Central Intelligence Agency. 1972. "Liberia- Sierra Leone Boundary." Washington, DC: Central Intelligence Agency. 1 map, color, 24 x 18 cm. Scale 1:1,160,000. Notes: Sources of data: Office of the Geographer, Department of State." Includes notes and key map. "501265 8-72." LCCN: 84-692339.

United States. Central Intelligence Agency. 1971. Liberia. Washington, D.C: Central Intelligence Agency. 1 map: color; 39 x 37 cm. Scale 1: 1,350,000. Notes: "Base 78456 2-71." "Shows boundaries, capitals and other populated places, roads, railroads, airfields, and chief port." LCCN: GM72-55.

United States. Central Intelligence Agency. 1971. Liberia. Washington, D.C: Central Intelligence Agency. 1 map: color; 39 x 37 cm on sheet 41 x 58 cm.. Scale 1: 1,350,000. Notes: "Relief shown by shading and spot heights." "Base 78455 2-71." "Margin includes location map, comparative area map, and maps of 'economic activity,' 'vegetation,' and 'population and tribal groups.'" LCCN: GM72-54.

United States. Central Intelligence Agency. 1963. Liberia. Washington, D.C: Central Intelligence Agency. 1 map: color; 35 x 37 cm. Scale 1: 1,350,000. Notes: "Relief shown by shading and spot heights." "10-63" LCCN: 96-687131.

United States. Central Intelligence Agency. 1959. Liberia. Washington, D.C: Central Intelligence Agency. 1 map: color; 20 x 21 cm. Scale 1: 2,400,000. LCCN: 96-680087.

United States. Central Intelligence Agency. 1955. West Africa-II (Southern Part) and Liberia. Section 23. Weather and Climate. Washington, DC.: The Agency. Descriptive Note: National intelligence survey. March 1955. 48 pages. Report number: NIS-50-II, NIS-51. Descriptors: climate; Subsaharan Africa; Nigeria; Ghana; Ivory Coast; weather; military operations; meteorological data. Identifiers: U/A reports, Dahomey, Togo, Liberia. Limitation Code: Approved for Public Release. DTIC Number: ADA950808; See also: AD-A950807.

United States. Coast and Geodetic Survey. 1955-1956. Liberia. Planimetric Maps. Coast and Geodetic Survey, Washington, DC. Scale: col. maps; 77 x 130 cm. or smaller. 1:125,000.

Notes: Coverage complete in 10 sheets. "Plane coordinates on Hotines rectified skew orthomorphic projection." Relief shown by hachures. LCCN: gm 71-2710; OCLC: 5567145.

United States. Coast and Geodetic Survey. 1957. Liberia: planimetric map. Washington: U.S. Coast and Geodetic Survey. 1 map on 2 sheets ; 82 x 58 cm. Scale 1:500,000. OCLC: 54325777.

United State. Department of State. Office of the Geographer. Map of Liberia. Np. 1941. Map Series 9. Publication 1623. 1 sheet in color, shows mountains by form lines. Index of geographic names. Inserts: Relative reliability; tribal distribution; physiography; boundary history. Scale 1:1,000,000. Held by USGS Library.

United States. Dept. of State. Division of Map Intelligence and Cartography. 1946. "Liberia: Principal Resources." Department of State, Division of Map Intelligence and Cartography. Washington, D.C.: The Division, [1946]. Cartographic Material. Description: 1 map: color; 24 x 25 cm. Scale Information: Scale 1:2,000,000 (W 12°--W 7°/N 9°--N 4°). Notes: Relief shown pictorially. "July 1946." "10423." Subjects: Liberia--Economic conditions--Maps. Library of Congress. Call Number: G8881.G1 1946 .U5. LC Control Number: 97-680064

United States Economic Mission to Liberia. 1962. "Republic of Liberia." Monrovia: The Mission. Topo map, 1:500,000. Notes: "Drawn by Dozer. 1 sheet, photocopy; 18 x 18 cm. Relief shown by hachures. LCCN: 96-687119.

United States Economic Mission to Liberia. 1951. "Republic of Liberia." Monrovia: The Mission. Topo map, 1:500,000. Notes: "Drawn by Rives; traced by Brunson. 1 sheet, photocopy; 68 x 117 cm.. Relief shown by hachures. Oriented with north toward upper left. The Library of Congress copy is torn along fold lines and along the edges of the sheet. LCCN: 97-680081.

United States Economic Mission to Liberia. 1949. "Republic of Liberia." Monrovia: The Mission. Topo map, 1:500,000. Notes: "Drawn by Rives; traced by Brunson. 1 sheet, photocopy; 68 x 117 cm.. Relief shown by hachures. Oriented with north toward upper left. Blue line print. LCCN: 95-680006.

United States. Economic Mission to Liberia. 1947. An introductory study of the soils of Liberia. Washington, D.C., Dept. of State. Description: 10, [4] l. tables. 28 cm. Subjects: Soils- Liberia. Notes: Typed. With this in the folder are: Agricultural resources, Republic of Liberia. Washington, D.C., Dept. of State - Forest resources, Republic of Liberia. Washington, D. C., Dept. of State [1947]. OCLC: 3577797.

United States Economic Mission to Liberia. 1949. "Republic of Liberia- Preliminary." Monrovia. Topo map, 1:500,000. Notes: 1 sheet, ozalid print. Shows mountains by "wooly caterpillars." Held in USGS Library.

United States. Geographic Names Division, Army Map Service. 1968. Liberia: Official Standard Names Approved by the United States Board on Geographic Names. Washington, DC. GPO 1968. iii, 61 pages, map. Series: Gazetteer. Number 106. "This gazetteer contains about 4,300 entries for places and features in Liberia. The entries consist of standard names approved by the Board on

Geographic Names and unapproved variant names, the latter cross-referenced to the standard names. Users of the gazetteer should always refer to the main entries for approved names. The density of name coverage corresponds to that of maps at the scale of 1:25,000. Entries include the names of first order administrative divisions, populated places of all sizes, various other cultural entities, and a variety of physical features." LCCN: 68-62724; USGS Library- Call number: 506 Un28g No.106.

United States. Geological Survey. 1968. "Liberia." In: Geological Survey Research for 1966. Chapter A. US Geological survey Professional Paper 550-A. Page 102.

United States. Office of Strategic Services. Geography Division. 1942. Liberia, roads and trails. Washington, D.C.: The Division, Edition: Provisional ed. Description: 1 map; 30 x 31 cm.; Scale [ca. 1:1,650,000]. Series: Map; no. 698. Subjects: Liberia- Road maps. Notes: Includes legend. Drawn in the Geography Division, O.S.S. OCLC: 38736138.

Utter, Thomas. 1993. "Gold mining potential of West Africa."Erzmetall. 46; 10, Pages 563-572. 1993. Dr. Riederer-Verlag. Stuttgart, Federal Republic of Germany. 1993. Language: English; Summary Language: German. Descriptors: Africa; Burkina-Faso; Ghana; gold ores; Guinea; Ivory Coast; Liberia; Mali; metal ores; metallogeny; mineral-deposits,-genesis; mineral exploration; production; Senegal; Sierra Leone; West Africa; Economic geology; geology of ore deposits. References: 12; illustrations, including portraits, 1 table, geological sketch maps. Reference includes data from Geoline, Bundesanstalt für Geowissenschaften und Rohstoffe, Hanover, Germany. ISSN: 0044-2658.

"Various Geologists [sic]." 1986. "Gold Mineralization in the Tortor Mountain Area: A Proposal to the United Nations." Liberia Geological Survey unpublished un-numbered report, 7 pages. AMRS, Inc. See: http://www.africaminerals.com/

Varnie, Josep Natanael, II. 2004. "Wealth extraction, not economic development: A case study on Liberia." 123 pages; [M.A. dissertation].United States, Massachusetts: University of Massachusetts Lowell; 2004. Abstract: Liberia, like many other third world countries, has not been able to attain any significant level of sustainable economic development since she declared her independence in 1847. In this thesis I discuss problems associated with the country's lack of sustainable development even though it is endowed with natural resources and fertile soil. Liberia had the potential to develop programs that would have resulted in a sustainable economy but because almost all of the country's resources were mortgaged to foreign concessions in the 1900s to maintain its independence it lacked the financial capacity to do so. To provide an historical account of what led to the country's prevailing economic situation, this research reviews four time periods: late 19th century during which the settlers arrived, the 1900s to 1940, the 1940s through the 1960s and the 1970s through early 1980s. This research pays particular attention to the 1940-1960 period when there was a major investment rush in Liberia by companies in extractive industries. Gross domestic income more than quadrupled, revenue receipts rose more than eightfold, the volume of imported goods nearly quadrupled, rubber exports rose by one-third from an already high export base, iron ore exports increased from nothing to nearly three million long tons per year, and labor market demand nearly tripled. Despite this economic expansion the vast majority of Liberians remained poor and lacked access to health facilities, schools, and safe drinking water. This research discusses the underlying factors that are responsible for the country's backwardness and lack of sustainability.

Venkatakrishnan R. and Culver S. 1989. "Tectonic fabric of Sierra Leone, West Africa: implications for Mesozoic continental breakup." Journal of the Geological Society, Volume 146, No. 6, 1989, pp. 991-1002(12). Abstract: A lineament map interpreted from Landsat images has been integrated with available geological and geophysical data from both offshore and onshore regions of Sierra Leone. The lineament patterns are related to intraplate and plate marginal reaction of pre-existing structures during Mesozoic rifting events that resulted in strong tectonic controls on magmatism. Of four main lineament trends, the NNW-SSE to N–S, NNE–SSW, and ENE–WSW trends are directly relatable to Archaean fabric in the Leo Uplift. A NW–SE trend reflects coast-parallel late Mesozoic dykes that follow the Rokelide Pan–African fabric (reactivated Archaean NNW and N–S trending structures). NW–SE trending faults defining offshore basins are segmented and offset by ENE–WSW trending continental extensions of ocean fracture zones. Both the Guinea and Sierra Leone Fracture Zones have nucleated on ENE–WSW trending sinistral Archaean shear zones in the Leo Uplift. The four lineament trends focused Mesozoic magmatic events through protracted reactivation. Spatial and geometric relationships between the magmatic provinces and tectonic fabric indicate that intraplate deformation occurs far inland during rifting events. The angular relationships between the Permo–Triassic NE trending Guinea Belt, the Jurassic–Triassic NW–SE trending coast-parallel dykes, and the ENE–WSW trending ocean fracture zones centred on the early Jurassic Freetown basic igneous complex, suggest that the Sierra Leone–Liberia continental margin evolved as an obliquely-sheared, rift–rift–transform passive margin during Mesozoic continental breakup.

Villeneuve, M. 1993. "The West African fold belts; structure and evolution." Comptes Rendus de l'Academie des Sciences, Serie 2, Mecanique, Physique, Chimie, Sciences de l'Univers, Sciences de la Terre. 316; 3, Pages 411-417. 1993. Gauthier-Villars. Montrouge, France. 1993. Language: English; Summary Language: French. Abstract: Before 1984, only two main orogenic periods had been considered in the West African fold belts (the first of Pan-African or Caledonian age, the second of Hercynian age). The discovery of two different Pan-African orogens ([I], [II]) partially reworked by a Hercynian tectonic event, was a substantial change in the interpretation of this fold belt. The first orogenic stage (Pan-African I) occurred in the northern part of the area around the Senegalese block; meanwhile, the second one (Pan-African II) occurred in the southern part, around the Rokelide block. The third stage (Hercynian) occurred only around the Senegalese block. The Pan-African I was first identified in the Bassaride ridge (South East Senegal), the Pan-African II in the Rokelide belt (Sierra Leone and Guinea) while the Hercynian has strongly reworked the Mauritanide belt. Each has different geodynamical patterns and geologic evolution despite their appearance as a single belt from Mauritania to Liberia. Descriptors: Africa; Caledonian-Orogeny; faults; Guinea-Bissau; Liberia; Mauritania; metamorphism; ophiolite; orogenic-belts; orogeny; Paleozoic; Pan-African-Orogeny; plate-collision; plate-tectonics; Precambrian; Proterozoic; rifting; sedimentation; Senegal; strike-slip-faults; suture-zones; tectonics; thrust-faults; upper Precambrian; West Africa; West African-Shield; Structural-geology. Illustrations: References: 26; illustrations, ISSN: 0764-4450.

Vladimirov, Boris Mikhaylovich; Tverdokhlebov, Viktor Aleksandrovich and Kolesnikova, Tamara Pavlovna (Tverdokhlebov, Viktor Aleksandrovich. [from old catalog] Kolesnikova,

Tamara Pavlovna. [from old catalog]). 1971. Geologiya i petrografiya izverzhennykh porod yugo-zapadnoy chasti Gvineysko-Liberiyskogo shchita. Translated title: Geology and petrography of igneous rocks of the southwestern part of the Guinea-Liberian shield. Akad. Nauk SSSR, Sib. Otd., Inst. Zemnoy Kory. Pages: 242. 1971. Description: 242 p. with illustrations, and maps. 21 cm. Language: Russian. Abstract: Rock types (Paleozoic peridotite-norite intrusives, Permo-Triassic trap rock, alkalic gabbroic rocks, Cretaceous kimberlites and other alkalic and ultramafic rocks), chemical analyses, structural setting, magmatism, tectonic relationships, west Africa. Descriptors: Africa; differentiation; genesis; Guinea Liberian Shield; igneous rocks; intrusions; magmas; Mesozoic; Paleozoic; petrology; processes; volcanism; volcanology; west; Igneous and metamorphic petrology; rocks, igneous; petrology, Guinea; [from old catalog] Petrology, Liberia. Illustrations, including sketch maps, Moscow. LCCN: 72-324520; USGS Library.

Vleechdraeger, Ed I. 1938. "Souvey [sic] of Road from Memehtown to Salala; with hand level and Compass." 1 map. Scale ½ = 5000; 1 cm=100 meters. Notes: "Signed, Ed. I. Vleechdraeger, Resident Engineer, Liberia." "Memeh Town May 11, 1938." Library of Congress.

Vogel, James William. 1982. "Late Quaternary sedimentary facies of the southern Sierra Leone and Liberian continental shelf and upper slope, Northwest Africa." Doctoral thesis, University of Rhode Island. Kingston, RI, United States. Pages: 375. 1982. Descriptors: Africa; areal studies; barrier islands; biogenic structures; bioturbation; Cenozoic; clastic sediments; clay; clay mineralogy; continental shelf; continental slope; currents; distribution; geophysical methods; geophysical surveys; Holocene; Liberia; lithofacies; marine sediments; marine transport; mud; oceanography; paleocurrents; Pleistocene; provenance; Quaternary; sedimentary structures; sedimentation; sediments; seismic methods; Sierra Leone; silt; surveys; thickness; transport; West Africa. Abstract (Document Summary): This study had three principal goals. One was to understand sedimentary processes on the continental margin off Sierra Leone and Liberia. This encompassed a study of late Quaternary sediments and included studies of geomorphology, bottom sediment characteristics, sediment distribution, provenance, facies, and environments; results then could be compared with hydrographic and related data. A second goal was an investigation of uncommonly extensive silt and clay shelf deposits, and a third was to detail geomorphic and geologic history. Sedimentologic, seismic, and bathymetric studies revealed a near-shore sand shoal, sandy shelf beds, thick silt and clay buildups comprising eight shelf lenses, and slope silts, clays, and canyon sands. A thin mud coating also covers much of the shelf and slope. The shoal comprises modified Pleistocene barriers. The sandy shelf contains Pleistocene to Holocene non-marine to inner neritic sands; outer shelf ridge- sets are drowned barrier island complexes. Shelf silt and clay lenses contain Holocene to modern suspension-derived sediments suggesting future hydrocarbon source potential. Slope deposits are Pleistocene and younger: sands represent suspensions with rolled grains and are turbidites, whereas slope silts and clays were deposited from suspensions by currents. The slope and canyons probably contain sediments re-suspended by shelf-break processes. The fine- clastic bottom coating appears related to a nepheloid layer. Silts and clays moved offshore from weathered basement along rivers showing abundant discharge related to monsoonal rains. Pellets and flocs aided deposition in lenses; bioturbation promoted bottom scour. Currents, swells, and seaward-diminishing sediment confined lenses mostly to inner shelf locations. Current-meters suggest waters probably move northwestward along shelf. Silt and clay distributions and grain-size parameters, however, indicate definite net northwest sediment transport. Lens isopachs reflect paleo-currents also with net northwest along shelf movements, indicating similar currents from Holocene to modern times. Offshore movements also occur. The combination of climate, provenance, drainage patterns, and sedimentary processes resulting in extensive silt and clay shelf bodies suggests a unique occurrence. This West African area might represent a new class of shelf sedimentation. Shoals formed during the Pleistocene. Regression caused sandy shelf beds. During the Holocene, barrier islands, beach or inner neritic sands, and silts and clays were deposited.

Vogelsang, Elke and Sarnthein, Michael. 2001. "Age control of sediment core GIK16776-1." PANGAEA. Note: Vogelsang, Elke; Sarnthein, Michael; Pflaumann, Uwe (2001): Stratigraphy, chronology, and sea surface temperatures of Atlantic sediment records (GLAMAP-2000 Kiel), Berichte-Reports, Geologisch-Paläontologisches Institut und Museum, Christian Albrechts Universität, Kiel, 13, 11 pages. See: Hüls, Matthias (1991): Meeresoberflächentemperaturen im Atlantik vor Liberia in den letzten 400.00 Jahren (Meteor Kern 16776), Diploma Thesis, Geologisch-Paläontologisches Institut, Christian Albrechts Universität, Kiel, 77 pages. Subjects: Age, comment; Age, dated; Age comment; Age dated; Age model; calculated; Flux-accumulation rates, geologic; Geology; Glacial Atlantic Mapping and Prediction; GLAMAP-2000; Institute for Geosciences, Christian Albrechts University, Kiel; M6/5; Meteor (1986); off Liberia; Piston corer; Piston corer (Kiel type); Sedimentation rate. Subjects: West: -11.3983; East: -11.3983; South: 3.7350; North: 3.7350; Minimum Depth, Sediment: 0.0 M. Maximum Depth, Sediment: 1.0 m. PANGAEA: Publishing Network for Geoscientific and Environmental Data. URL: http://doi.pangaea.de/10.1594/PANGAEA.59670

Voigt, Fritz. and Heuskel, Dieter. 1981. Sozioökonomische Folgewirkungen privatwirtschaftlicher Direktinvestitionen in Entwicklungsländern: dargestellt am Beispiel der Bong Mining Company in Liberia. Berlin: Duncker & Humblot. 230 pages; 21 cm. Language: German. Schriftenreihe zur Industrie- und Entwicklungspolitik; Bd. 25. Descriptors: Investments, Foreign- Liberia; Investments, Foreign – Developing countries- Case studies; Liberia- Economic conditions- 1971-1980. Named Corp: Bong Mining Company. Notes: Bibliography: p. 219-230. ISBN: 3428048601; LCCN: 81-159515; OCLC: 8113717.

Volz, Walter, 1875-1907. 1911. Reise durch das Hinterland von Liberia im Winter 1906-1907; Nach seinen Tagebeuchern bearbeitet von Dr. Rudolf Zeller. Translated "Journey to the hinterland of Liberia in the winter of 1906-1907; After the day books of Dr. Rudolf Zeller." Bern: Francke, 1911. Description: Book. 167 pages: illus., front. (port.), maps (fold.) ; 24 cm. Notes: Includes bibliographical references. Descriptors: Liberia Description and travel; Africa, Western Description and travel. Other: Zeller, Rudolf. University of Wisconsin, American Geographical Society Library. Call Number: DT626 .V65x 1911.

Wallace, R. M. 1977. "Geologic map of the Bopolu Quadrangle, Liberia." Miscellaneous Investigations Series Map. Report number: I-0772-D. January 1, 1977. U. S. Geological Survey, Reston, VA, United States. Map Type: geologic map; color map; 44 x 63 cm. on sheet 74 x 106 cm. folded in envelope 30 x 24 cm. Map Scale: 1:250,000 (1 inch = about 4 miles). Sheet 29 by 42 inches. Descriptors: Africa; areal geology; Bopolu Quadrangle; geologic maps; Liberia; maps; USGS; West Africa. Notes: "Hotines rectified skew orthomorphic projection and rectangular coordinates." "Prepared by the U.S. Geological Survey and Liberian Geological Survey under the joint sponsorship of the Government of Liberia and the Agency for International Development, U.S. Department of State." Includes bibliography. ISSN: 0160-0753.

Wallace, R. M. 1974. "Geology of the Bopolu Quadrangle, Liberia." Open-File Report. Report number: OF 74-0302. Pages (monograph): 13 pages (1 sheet). January 1, 1974. U. S. Geological Survey, Reston, VA, United States. Annotation: 1 sheet, scale 1:250,000 (1 inch = about 4 miles). Descriptors: Africa; areal geology; Bopolu Quadrangle; Liberia; maps; USGS; West Africa. Notes: Map in pocket. Liberian investigations, Project report (IR) LI-61B. Prepared under the auspices of Govt. of Liberia and Agency for International Development. Letter of transmittal dated December 13, 1974. Bibliography: leaf 13. ISSN: 0196-1497.

Wallis, C. Braithwaite, Captain. 1909. "Prismatic Compass Traverse in Liberia." Scale 1:500,000, or one inch=7.89 statute miles. Insert from the Geographical Journal. 1910. Notes: C. B. Wallis, "H. B. M. Consul-General at Dakar, lately H. B. M. Consul-General for Liberia." "Boundaries of Chiefdoms." "Stockaded Villages." Library of Congress, Geography & Map Division.

West Africa-II (Southern Part) and Liberia. Section 23. Weather and Climate. 1955. From Independent Federal Agencies. Report Number: NIS-50-II. March 1955. 48 Pages(s) Descriptive Note: National intelligence survey. Descriptors: Climate; Subsaharan Africa; Nigeria; Ghana; Ivory Coast; weather; military operations; meteorological data. Approved For Public Release. DTIC: ADA950808; See also AD-A950 807.

Westerhausen, Lothar. 2003. "Organic chemistry analysis on surface sediments from the equatorial east Atlantic." PANGAEA. Note: Westerhausen, Lothar (1992): Organische Sedimente im äquatorialen Ostatlantik: Einflüsse von Herkunft, Transportmustern, Diagenese und Klimaschwankungen, Berichte-Reports, Geologisch-Paläontologisches Institut und Museum, Christian Albrechts Universität, Kiel, 48, 109 pages. Subjects: Alkane/total organic carbon ratio; Alkanes; ANT-IV/1c; Atlantic Ocean; Box corer/grab; C; C/N; calculated; Carbon, organic, terrestrial matter; Carbon, organic, total; Carbon, total; Carbon/Nitrogen ratio; Chemistry, organic compounds; Chemistry, sediment; Components, terrigeneous; d13C Corg; delta 13C, organic carbon; Dinost/TOC; Dinosterol/total organic carbon ratio; Dust; Dust, aeolian; eastern Romanche Fracture Zone; Element analyser CHN; Element analyser CNS, Carlo Erba NA1500; Equatorial Atlantic; Gas volumetric; Giant box corer; Gravity corer; Gravity corer (Kiel type); Higher Plant Alkanes index; HPA-index; Institute for Geosciences, Christian Albrechts University, Kiel; Isotopes, stable, general; Ketone/TOC; M6/5; M65; Mass spectrometer Finnigan MAT 251; Meteor (1964); Meteor (1986); NE Atlantic off Liberia; normalized; Oceanography; off Cote d Ivoire; off eastern Ghana; off Gabun; off Ghana; off Guinea; off Lagos; off Liberia; off Nigeria; off Nigeria-Delta; Piston corer; Piston corer (Kiel type); Polarstern; Sierra Leone Basin/Guinea Basin. Subjects: West: -22.8650; East: 9.0167; South: -2.2033; North: 12.2600. Minimum Depth, Sediment: 0.0 M. Maximum Depth, sediment: 0.0 m. PANGAEA: Publishing Network for Geoscientific and Environmental Data. URL: http://doi.pangaea.de/10.1594/PANGAEA.89382

Westerhausen, L.; Poynter, J.; Eglinton, G.; Erlenkeuser, H. and Sarnthein, M. 1993. "Marine and terrigenous origin of organic matter in modern sediments of the equatorial East Atlantic; the delta (super 13) C and molecular record." Deep-Sea Research. Part I: Oceanographic Research Papers. 40; 5, Pages 1087-1121. 1993. Pergamon. Oxford-New York, International. 1993. Descriptors: Africa; aliphatic-hydrocarbons; alkanes; Atlantic Ocean; C-13-C-12; carbon; Central Africa; continental shelf; continental slope; East Atlantic; Gabon; hydrocarbons; isotope ratios; isotopes;

Ivory Coast; Liberia; marine sedimentation; molecular structure; organic carbon; organic compounds; organic materials; sedimentation; stable isotopes; temperature; terrestrial environment; West Africa; isotope geochemistry; oceanography. Illustrations: References: 99; illustrations, including 2 tables, geological sketch maps. ISSN: 0967-0637.

White, Lane. 1985. "Bong iron ore: rationalization improves efficiencies at Liberian operation." Engineering and Mining Journal v. 186 (December 1985, pages 24-8. Descriptors: Iron mines and mining- Liberia; Mining methods- Stripping operations; Iron ores- Pelleting. ISSN: 0095-8948.

White, Lane. 1985. "Liberian iron ore; rebounding LAMCO now seeks a long term future." E and MJ, Engineering and Mining Journal. 186; 10, Pages 25-30. 1985. McGraw Hill. New York, NY. Descriptors: Africa; chemically precipitated rocks; development; economic geology; Guinea; iron formations; iron ores; Liberia; metal ores; Mifergui ine; Nimba Mountains; production; reserves; sedimentary rocks; West Africa. Illustrations, including 4 tables, sketch map. ISSN: 0095-8948.

White, R. W. 1973. "Progressive metamorphism of iron formation and associated rocks in the Wologizi Range, Liberia." USGS Bulletin. Report number: B 1302. January 1, 1973. (other date): January 1, 1974. Pages: 50. Maps (1 folded color in pocket); 24 cm. U. S. Geological Survey, Reston, VA, United States. Map Scale: 1:100,000; colored geologic map. Annotation: 1 plate in pocket. Illustrations, including tables. Descriptors: Africa; chemical composition; chemically precipitated rocks; geologic; grade; iron formations; iron ores; Liberia; maps; metal ores; metamorphic rocks; metamorphism; metasedimentary; mineral assemblages; ore deposits; petrology; possibilities; sedimentary rocks; USGS; variations; weathering; West Africa; Wologizi Range; zoning. Subjects: Metamorphism (Geology). Petrology- Liberia. Notes: Bibliography: p. 49-50. ISSN: 8755-531X; LCCN: 73-600266; OCLC: 1253863.

White, Richard W. 1972. "Progressive metamorphism of iron formation and associated rocks in the Wologizi Range, Liberia." Open-file report (Geological Survey (U.S.)); 1765. Washington: U.S. Geological Survey, 1972. 90 leaves: ill., maps; 28 cm. Notes: Two maps folded in pocket. Prepared under the auspices of the Govt. of Liberia and the Agency for International Development, U.S. Dept. of State. Referred to in press release dated August 18, 1972. Bibliography: leaves 87-90. Descriptors: iron ores- Liberia; Wologizi Range; rocks, metamorphic. USGS Library.

White, Richard W. 1972. "Progressive metamorphism of iron formation and associated rocks in the Wologizi Range, Liberia." Open-File Report. Report number: OF 72-0450. Pages (monograph): 90 pages (2 sheets). Date: January 1, 1972. U. S. Geological Survey, Reston, VA, United States. Annotation: 2 sheets. Descriptors: Africa; economic geology; iron formation; iron ores; Liberia; metal ores; metamorphism; petrology; prograde metamorphism; progressive metamorphism; rates; rocks; USGS; West Africa; Wologizi Range. ISSN: 0196-1497.

White, Richard W. 1972. Stratigraphy and structure of basins on the coast of Liberia. Monrovia: Republic of Liberia, Ministry of Lands and Mines, Liberian Geological Survey, Special Papers - Liberia, Geological Survey. 3; 1972. 14 p.: ill., maps (some folded in pocket). Notes: Bibliography: p. 14. Map scale: 1:250,000. Descriptors: Africa; areal geology; basins; coastal; Cretaceous; deposition; Dip slip; displacements; economic geology; faults; geologic; Liberia; Liberia, Physical geology; lithostratigraphy; maps; Mesozoic; offshore; petroleum; possibilities; sedimentary petrology; sedimentation; stratigraphy; structural geology; West Africa; Geologic maps. ISSN: 0375-6831; LCCN: 76-359122; OCLC: 2931231; 1477134.

White, Richard W. et al. 1970. "Map- Preliminary geologic map of western Liberia." US Geological Survey. Washington, DC. Notes: Plate 1 of US Geological Survey Special paper 1. Scale 1:1,000,000. 1 sheet in color. Held in USGS Library Map Collection.

White, Richard William. 1970. "Reconnaissance geologic mapping in Liberia." Washington: U.S. Geological Survey. 47 leaves: ill., maps; 27 cm. Open-file report (Geological Survey (U.S.)); 1358. Notes: Prepared as part of the joint Geological Exploration and Resources Appraisal project of the Liberian Geological Survey sponsored by the Government of Liberia and the United States Agency for International Development. Referred to in press release dated February 16, 1970. Bibliography: leaves 45-47. USGS LIBRARY.

White, Richard William. 1970. "Reconnaissance mapping of deeply weathered crystalline rocks in Liberia." Bulletin - Geological, Mining and Metallurgical Society of Liberia. 4; Pages 1-25. 1970. Notes: Vol. 4. Geological, Mining and Metallurgical Society. Monrovia, Liberia. Descriptors: Africa; granites; igneous rocks; Liberia; metamorphic rocks; petrology; plutonic rocks; West Africa. Illustrations, including sketch maps. ISSN: 0367-4819.

White, Richard William. 1969. "Sedimentary rocks of the coast of Liberia." Washington: U.S. Geological Survey; Liberian Geological Survey.; United States. Agency for International Development.. 35 leaves: ill., maps; 27 cm. Series: Open-file report (Geological Survey (U.S.)); 1322. Subjects: Geology- Liberia. Notes: Project report IR-LI-39. Includes bibliographical references (leaf 34-35). Photocopy. [S.l.: s.n., 19--]. 28 cm. Prepared as part of the joint Geological Exploration and Resources Appraisal project of the Liberian Geological Survey and the U.S. Geological Survey sponsored by the Government of Liberia and the United States Agency for International Development. Referred to in press release dated October 31, 1969. Bibliography: leaf 34-35. USGS Library; OCLC: 53141397.

White, Richard W. and Leo, Gerhard W. 1970. "Geologic summary of age provinces in Liberia." Bulletin - Geological, Mining and Metallurgical Society of Liberia. 4; Pages 96-102. 1970. Notes: Vol. 4. Geological, Mining and Metallurgical Society. Monrovia, Liberia. Descriptors: absolute age; Africa; dates; geochronology; Liberia; West Africa. Illustrations: sketch map. ISSN: 0367-4819.

White, Richard W. and Leo, Gerhard W. 1969. Geologic reconnaissance in western Liberia. Special Papers - Liberia, Geological Survey. 1; 1969. Monrovia, Liberia. Pages: 18. Abstract: Description and structure of metamorphic, igneous, and sedimentary rocks, geochronology, regional correlations. Descriptors: absolute age; Africa; areal geology; basalts; dates; diabase; geochronology; geologic; igneous rocks; K-Ar; Liberia; maps; metamorphic rocks; petrology; plutonic-rocks; Rb-Sr; stratigraphy; volcanic rocks; west; West Africa; Geologic maps. MAP SCALE: 1:1,000,000. Illustrations, color geologic maps. ISSN: 0375-6831.

White, Richard W. and Leo, Gerhard W. 1968. Preliminary geologic map of western Liberia. Monrovia?; Liberian Geological Survey,]; Interior- Geological Survey. Description: 1 map: color;

42 x 28 cm. Subjects: Geology- Liberia- Maps. Map Info: Scale 1:1,000,000; (W 11030'—W 9000'/N 8030'--N 5000'). Notes: "Geological reconnaissance in western Liberia, special paper 1, plate 1." "Prepared as part of the joint geological exploration appraisal project of the Liberian Geological Survey and the U.S. Geological Survey sponsored by the government of Liberia and the United States Agency for International Development." Includes index to geologic mapping.

Williams, H. R. and Culver, S. J. 1982. "The Rokelides of West Africa; Pan-African aulacogen or back-arc basin?" Precambrian Research. 18; 3, Pages 261-273. 1982. Elsevier. Amsterdam, International. Descriptors: Africa; aulacogens; basins; Guinea; Guinea Bissau; Liberia; marginal basins; mechanism; orogeny; Pan African Orogeny; Precambrian; Proterozoic; Rokelides belt; Senegal; Sierra Leone; stratigraphy; structural geology; tectonics; upper Precambrian; West Africa; Structural geology. Illustrations, including sect., strat. color, geological sketch map. ISSN: 0301-9268.

Woiwor, Charles Dresser. 1990. Alternative fiscal regimes applicable to the Mifergui Joint Project. Description: ix, 94 leaves: ill., map; 28 cm. Descriptors: investments, foreign- taxation- Guinea; investments, foreign- taxation- Liberia; taxation- Guinea; taxation- Liberia; iron mines and mining-Guinea; iron industry and trade- Guinea; iron industry and trade- Liberia. Named Corp: Mifergui Joint Project. Notes: Typescript (photocopy). Includes bibliographical references (leaves 76-78). Dissertation: Thesis (M. Sc.)--Colorado School of Mines. OCLC: 23113464.

Wood, D. A. 1982. "A regional study of the potential for hydrocarbon exploration in Africa: including map showing the distribution and geological characteristics of the sedimentary basins." Phillips Petroleum Company Europe-Africa.; Africa Exploration. London: Phillips Petroleum Co. Europe-Africa. 1 volume (loose-leaf); 29 cm. Descriptors: Petroleum- Geology – Africa; Petroleum- Geology- Liberia. OCLC: 14223293.

Worrall, George Alan. 1969. "Present-day and sub fossil beach cusps on the west African coast." Journal of Geology. 77; 4, Pages 484-487. 1969. University of Chicago Press. Chicago, IL, United States. Abstract: Environment and mechanism of formation, less transient feature than generally supposed, sub fossil occurrences in former beach ridges now inland, Sierra Leone, Liberia. Descriptors: Africa; beach cusps; coast; geomorphology; Liberia; shore features; Sierra Leone; West Africa; geomorphology. Illustrations. ISSN: 0022-1376.

Worrall, George Alan. 1967. Agricultural research in Liberia : a review and summary of publications by G. A. Worrall. Rome: Food and Agriculture Organization of the United Nations, 1967. 85 pages; 29 cm. Notes: "RU: SF/67/4" Descriptors: agricultural literature Liberia; agriculture Liberia Bibliography. Other: Food and Agriculture Organization of the United Nations. University of Wisconsin- American Geographical Society Collection. Call Number: S473.L7 W67 1967

Worthington, E. B. 1939. Science in Africa. New York: Oxford University Press. 746 pages. Maps. Abstract: This book is one of a series of reports prepared in connection with the African Research Society. The problems of Africa, as they present themselves to those whose concern is with the development of the continent, are discussed in "An African Survey." The purpose of this volume is to summarize the present position of studies in the various sciences which have a bearing on African conditions.

Wotorson, C. S. and Behrendt, John Charles. 1974. "Total-count gamma radiation map of the Zwedru Quadrangle, Liberia." Other Titles: "Total count gamma radiation map of the Zwedru quadrangle, Liberia. Zwedru, Liberia, gamma radiation." Miscellaneous Investigations Series Map. Date: January 1, 1974. Report number: I-0777-C. U. S. Geological Survey, Reston, VA, United States. Map Type: geophysical survey map. 1 map: color; 45 x 55 cm. folded in envelope 30 x 24 cm. Annotation: Lat 6 degrees to 7 degrees, long 7 degrees to 9 degrees. Scale 1:250,000 (1 inch = about 4 miles). Sheet 28 by 32 inches. Descriptors: Africa; airborne; east; gamma ray; geophysical methods; geophysical surveys; Liberia; maps; radioactivity methods; surveys; USGS; West Africa; Zwedru Quadrangle. Notes: Relief shown by spot heights. Envelope Zwedru, Liberia, gamma radiation. "Prepared in cooperation with the Republic of Liberia, Bureau of Natural Resources and Surveys." "Prepared under the joint sponsorship of the government of Liberia and the Agency for International Development, U.S. Department of State." Includes text, inset map, 2 ancillary maps, and bibliography. Geologic maps. ISSN: 0160-0753.

Wotorson, Cletus S. and Behrendt, John Charles. 1971. "Aeromagnetic map of the Juazon quadrangle, Liberia." Washington: U.S. Geological Survey, 1971. 5 [i.e. 8] leaves: maps; 27 cm. Series: Open-file report (Geological Survey (U.S.)); 1597. Notes: At head of Project report, Liberian investigations (IR) LI-75 C. Some maps folded in pocket. Referred to in press release dated August 2, 1972. Bibliography: leaf 5. USGS Library.

Wotorson, C. S. and Behrendt, John Charles 1971. "Aeromagnetic map of the Juarzoa Quadrangle, Liberia." Open-File Report. Report number: OF 71-330. 5 pages, 2 sheets. U. S. Geological Survey, Reston, VA, United States. January 1, 1971. 1 map on 2 sheets text. Map Scale: 1:250,000. Annotation: 2 sheets, scale 1:250,000 (1 inch = about 4 miles). Descriptors: aeromagnetic maps; Africa; airborne methods; explanatory text; geophysical methods; geophysical surveys; Juarzoa Quadrangle; Liberia; magnetic methods; maps; surveys; USGS; West Africa. ISSN: 0196-1497.

Wotorson, Cletus S. and Behrendt, John Charles. 1974. "Aeromagnetic map of the Zwedru Quandrangle, Liberia." Miscellaneous Investigations Series Map. Date: January 1, 1974. Report number: I-0777-B. U. S. Geological Survey, Reston, VA, United States. Map Type: magnetic survey map. 1 map: color; 43 x 56 cm. folded in envelope 30 x 24 cm. Annotation: Lat 6 degrees to 7 degrees, long 7 degrees to 9 degrees. Scale 1:250,000 (1 inch = about 4 miles). Sheet 28 by 32 inches. Descriptors: Africa; airborne; east; geophysical methods; geophysical surveys; Liberia; magnetic methods; maps; surveys; USGS; West Africa; Zwedru Quadrangle. Notes: Relief shown by spot heights. Envelope Zwedru, Liberia, aeromagnetic. "Prepared in cooperation with the Republic of Liberia, Bureau of Natural Resources and Surveys." "Prepared under the joint sponsorship of the government of Liberia and the Agency for International Development, U.S. Department of State." Includes text, maps, and bibliography. Geologic maps. ISSN: 0160-0753; OCLC: 15324719.

Wotorson, C. S. and Behrendt, John Charles. 1971. "Aeromagnetic map of the Zwedru Quadrangle, Liberia." Open-File Report. Report number: OF 71-0331. 5 pages, 2 sheets. U. S. Geological Survey, Reston, VA, United States. January 1, 1971. Scale: 1:250,000. Map Type: aeromagnetic map. Annotation: 2 sheets, scale 1:250,000 (1 inch = about 4 miles). Descriptors: aeromagnetic maps; Africa; airborne methods; explanatory text; geophysical methods; geophysical surveys; Liberia; magnetic methods; maps; surveys; USGS; West Africa; Zwedru Quadrangle. ISSN: 0196-1497.

Wotorson, Cletus S. and Behrendt, John Charles. 1971. "Aeromagnetic map of the Zwedru quadrangle, Liberia." Washington: [U.S. Geological Survey], 1971. 5 leaves: maps (4 folded); 27 cm. Open-file report (Geological Survey (U.S.)); 1602. Notes: Project report, Liberian investigations (IR) LI-70 C. Referred to in press release dated August 2, 1971. Bibliography: leaf 5. USGS Library.

Wotorson, Cletus S. and Behrendt, John Charles. 1971. "Total-count gamma radiation map of the Zwedru quadrangle, Liberia." Washington: U.S Geological Survey, 1971. 5 leaves: maps; 27 cm. Series: Open-file report (Geological Survey (U.S.)); 1622. Notes: Four maps folded in pocket. Referred to in press release dated October 12, 1971. Bibliography: leaf 5. Descriptors: Gamma rays-Measurement. USGS Library.

Wotorson, C. S. and Behrendt, John Charles. 1971. "Total-count gamma radiation map of the Zwedru Quadrangle, Liberia." Open-File Report. Report number: OF 71-0332. Pages (monograph): 5 (2 sheets). January 1, 1971. U. S. Geological Survey, Reston, VA, United States. Map Scale: 1:250,000. Annotation: 2 sheets, scale 1:250,000 (1 inch = about 4 miles). Descriptors: Africa; gamma rays; geophysical methods; geophysical survey maps; geophysical surveys; Liberia; maps; radioactivity methods; surveys; USGS; West Africa; Zwedru Quadrangle, geophysical survey map. ISSN: 0196-1497.

Wright, J. B.; Hastings, D. A.; Jones, W. B. and Williams, H. R. 1988. Geology and Mineral Resources of West Africa. New York: Springer-Verlag. 176 pages. ISBN: 0045560013.

Youngman, E. P. 1932. Mining laws of the Republic of Liberia. Washington, D.C.: U.S. Dept. of the Interior, Bureau of Mines. Description: 11 pages. Series: Information circular. Bureau of Mines; 6630; Variation: Information circular (United States. Bureau of Mines); Number 6630. Subjects: Mining law- Liberia. Notes: "One of a series of digests of foreign mining laws." OCLC: 41652457.

Zakharov, V. A. 1979. "Die Zonenkorrelation des borealen Neokom mit Buchien." Translated title: "Zone correlation of the Boreal Neocomian with Buchia." In: Aspekte der Kreide Europas. Wiedmann, J (editor). International Union of Geological Sciences. Series A. 6, Pages 117-120. 1979. International Union of Geological Sciences. Stuttgart, Federal Republic of Germany. Conference: 1. Symposium Deutsche Kreide; Bindeglied zwischen Boreal und Tethys. Munster, Westphalia, Federal Republic of Germany. April 1978. Language: German; Summary Language: English. Descriptors: Ammonoidea; Angiospermae; Arctic region; Berriasian; biostratigraphy; Buchia; Cephalopoda; Commonwealth of Independent States; correlation; Cretaceous; Europe; Hauterivian; Invertebrata; Lower Cretaceous; Mesozoic; Mollusca; Neocomian; northern Liberia; Plantae; Russian Plain; Simbirskites; Spermatophyta; stratigraphy; Tetrabranchiata; USSR; Valanginian. References: 9; table. ISBN: 3510560043; ISSN: 0374-8480.

Zeba, S. 2005. "Community wildlife management in West Africa : a regional overview."
Note: This report is intended to be a West African contribution to a global study of IIED on
community wildlife management issues. Its geographic focus is the 16 member countries of the
Economic Community of West African States (ECOWAS), including 9 francophone countries
(Benin, Burkina Faso, Niger, Mali, Ivory-Coast, Mauritania, Senegal, Guinea,Togo), 5 anglophone
countries (Ghana, Liberia, Nigeria, Sierra Leone, The Gambia) and 2 lusophone countries (Guinea
Bissau, Cape Verde). This region has more than 200 million inhabitants. Eight (8) of the 16
countries concerned are part of the Sahelian region, and are members of the Permanent Interstate
Committee for Drought Control in the Sahel (CILSS). The remaining ones are generally considered
as being better endowed with natural resources (e.g. flora and fauna species, forests resources,
water, etc.) because of their location in a semi-forest zone. However, desertification (known as a
broad process of land degradation) has been reported to be affecting most forest countries also, and
this might be the reason why all the 15 other countries of the region, except Liberia, have ratified
the International Convention to Combat Desertification. It should be noted also that Benin has tried
to join CILSS these last years. Coastal erosion and deforestation are other serious problems
affecting those forest countries on the Atlantic Coast. The combined action of drought,
desertification, deforestation and population pressure, have widely depleted natural resources and
wildlife. Note: World Conservation Union, Foundation NATURAMA, IIED. Evaluating Eden
Series, Working Paper No.9. URL: http://www.iied.org/docs/blg/eden_dp9.pdf;
http://hdl.handle.net/1834/659.

Zech, W. and Drechsel, P. 1992. "Multiple mineral deficiencies in forest plantations in Liberia."
Forest Ecology & Management 48, no.1-2 (1992) p. 121-143. Abstract: Foliar analysis of Pinus
caribaea, P. oocarpa and several exotic as well as native broad-leaved species show obvious
differences in the nutrient status of fast-growing, healthy trees and slowly growing trees with
deficiency symptoms. Besides P, foliar levels of K, Mg and N seem to be below the critical range
in pines. On the other hand, high Al and Fe concentrations occur frequently. Descriptors: evolution
and palaeoecology; nutrient deficiency; Pinus oocarpa. ISSN: 0378-1127.

Zech W. and Drechsel, P. 1991. "Relationships between growth, mineral nutrition and site factors
of teak (Tectona grandis) plantations in the rainforest zone of Liberia." Forest Ecology &
Management 41, no.3-4 (1991) p. 221-235. Abstract: In 5-11 year-old teak plantations at Glaro,
Cavalla and Bomi Hills in Liberia, growth and vigour of trees show considerable variations.
Differences in growth intensity are mainly related to topsoil acidity and foliar Ca status. There are
also mineral disorders concerning the N, P and Mn supplies of teak. Descriptors: evolution and
palaeoecology; Tectona grandis. ISSN: 0378-1127.

Zech, W.; Haumaier, L. and Koegel, Knabner I. 1989. "Changes in aromaticity and carbon
distribution of soil organic matter due to pedogenesis." In: Advances in humic substances research;
a collection of papers from the Fourth international meeting of the International Humic Substances
Society. Saiz, Jimenez C. et al., eds. The Science of the Total Environment. 81-82; Pages 179-186.
1989. Elsevier. Amsterdam, Netherlands. 1989. Conference: International Humic Substances
Society, Fourth international meeting. Huelva, Spain. October 3-7, 1988. Descriptors: Africa; alkyl
carbon; Alps; aromaticity; Bavaria, Germany; Bavarian Alps; carbon; Central Alps; Central
Europe; cross polarization magic angle spinning technique; distribution; Europe; Fichtelgebirge;
Galicia Spain; geochemistry; Germany; Iberian Peninsula; Liberia; lignins; litter; NMR spectra;

organic compounds; organic materials; pedogenesis; soils; Southern Europe; Spain; spectra; West Africa; West Germany. Illustrations: References: 8; illustrations, including 1 table. ISSN: 0048-9697

Zepter, K. H. 1967. "Bong Range, Bergbaubetreib in einem Entwicklungsland." Translated title: "Mining industry- Operations in a Developing Country." Zeitschr. Erzbergbau u. Metallhüttenwesen. Volume 20, no. 1, pages 1-7.

Zinchuk, N. N. 1982. "Mineral composition of kelyphytic rims on garnets from kimberlites." International Geology Review. 24; 3, Pages 354-358. 1982. Winston & Son. Silver Spring, MD, United States. Descriptors: Africa; Asia; Commonwealth of Independent States; diamonds; economic geology; garnet group; Guinea; igneous rocks; intrusions; kimberlite; Liberia; Liberian Shield; mineral exploration; nesosilicates; ore guides; orthosilicates; pipes; plutonic rocks; reaction-rims; Russian Federation; Russian Republic; Siberia; silicates; ultramafics; USSR; West Africa; Yakutia Russian Federation; Yakutia Russian Republic. References: 15; illustrations, including 15 anal., 1 table. ISSN: 0020-6814.

Map Series:

The USGS Map Folios of Liberia, with a map scale of 1:250,000, are available from many libraries, or the US Geological Survey: http://www.usgs.gov/
Map I-771 A, 1973, Geographic Map of the Voinjama Quadrangle, Liberia
Map I-771 B, 1974, Aeromagnetic Map of the Voinjama Quadrangle, Liberia
Map I-771 C, 1974, Total-Count Gamma Radiation Map of the Voinjama Quadrangle, Liberia
Map I-771 D, 1974, Geologic Map of the Voinjama Quadrangle, Liberia
Map I-771 E, 1979, Mineral Localities of the Voinjama Quadrangle, Liberia
Map I-772 A, 1973, Geographic Map of the Bopolu Quadrangle, Liberia
Map I-772 B, 1974, Aeromagnetic Map of the Bopolu Quadrangle, Liberia
Map I-772 C, 1974, Total-Count Gamma Radiation Map of the Bopolu Quadrangle, Liberia
Map I-772 D, 1974, Geologic Map of the Bopolu Quadrangle, Liberia
Map I-772 E, 1979, Mineral Localities of the Bopolu Quadrangle, Liberia
Map I-773 A, 1973, Geographic Map of the Zorzor Quadrangle, Liberia
Map I-773 B, 1974, Aeromagnetic Map of the Zorzor Quadrangle, Liberia
Map I-773 C, 1974, Total-Count Gamma Radiation Map of the Zorzor Quadrangle, Liberia
Map I-773 D, 1974, Geologic Map of the Zorzor Quadrangle, Liberia
Map I-773 E, 1979, Mineral Localities of the Zorzor Quadrangle, Liberia
Map I-774 A, 1973, Geographic Map of the Sanokole Quadrangle, Liberia
Map I-774 B, 1974, Aeromagnetic Map of the Sanokole Quadrangle, Liberia
Map I-774 C, 1974, Total-Count Gamma Radiation Map of the Sanokole Quadrangle, Liberia
Map I-774 D, 1974, Geologic Map of the Sanokole Quadrangle, Liberia
Map I-774 E, 1979, Mineral Localities of the Sanokole Quadrangle, Liberia
Map I-775 A, 1973, Geographic Map of the Monrovia Quadrangle, Liberia
Map I-775 B, 1974, Aeromagnetic Map of the Monrovia Quadrangle, Liberia
Map I-775 C, 1974, Total-Count Gamma Radiation Map of the Monrovia Quadrangle, Liberia
Map I-775 D, 1974, Geologic Map of the Monrovia Quadrangle, Liberia
Map I-775 E, 1979, Mineral Localities of the Monrovia Quadrangle, Liberia
Map I-776 A, 1973, Geographic Map of the Gbanka Quadrangle, Liberia

Map I-776 B, 1974, Aeromagnetic Map of the Gbanka Quadrangle, Liberia
Map I-776 C, 1974, Total-Count Gamma Radiation Map of the Gbanka Quadrangle, Liberia
Map I-776 D, 1974, Geologic Map of the Gbanka Quadrangle, Liberia
Map I-776 E, 1979, Mineral Localities of the Gbanka Quadrangle, Liberia
Map I-777 A, 1973, Geographic Map of the Zwedru Quadrangle, Liberia
Map I-777 B, 1974, Aeromagnetic Map of the Zwedru Quadrangle, Liberia
Map I-777 C, 1974, Total-Count Gamma Radiation Map of the Zwedru Quadrangle, Liberia
Map I-777 D, 1974, Geologic Map of the Zwedru Quadrangle, Liberia
Map I-777 E, 1979, Mineral Localities of the Zwedru Quadrangle, Liberia
Map I-778 A, 1973, Geographic Map of the Buchanan Quadrangle, Liberia
Map I-778 B, 1974, Aeromagnetic Map of the Buchanan Quadrangle, Liberia
Map I-778 C, 1974, Total-Count Gamma Radiation Map of the Buchanan Quadrangle, Liberia
Map I-778 D, 1974, Geologic Map of the Buchanan Quadrangle, Liberia
Map I-778 E, 1979, Mineral Localities of the Buchanan Quadrangle, Liberia
Map I-779 A, 1973, Geographic Map of the Juazohn Quadrangle, Liberia
Map I-779 B, 1974, Aeromagnetic Map of the Juazohn Quadrangle, Liberia
Map I-779 C, 1974, Total-Count Gamma Radiation Map of the Juazohn Quadrangle, Liberia
Map I-779 D, 1974, Geologic Map of the Juazohn Quadrangle, Liberia
Map I-779 E, 1979, Mineral Localities of the Juazohn Quadrangle, Liberia
Map I-780 A, 1973, Geographic Map of the Harper Quadrangle, Liberia
Map I-780 B, 1974, Aeromagnetic Map of the Harper Quadrangle, Liberia
Map I-780 C, 1974, Total-Count Gamma Radiation Map of the Harper Quadrangle, Liberia
Map I-780 D, 1974, Geologic Map of the Harper Quadrangle, Liberia
Map I-780 E, 1979, Mineral Localities of the Harper Quadrangle, Liberia

Topographic Maps. This collection of maps is primarily photographic negatives of maps (for printing positive images of maps), 1:50,000 original scale. The UK-DOS (United Kingdon, Directorate of Overseas Surveys) and US-DMA (Defense Mapping Agency) sheets are also on file as a single hard copy colored map from which copies could also be made. Copies are available from many libraries, or from AMRS, Inc.

Map Name	Map Series	Sheet Number	Publisher	Date
Barziwen	Lib 50	2641/II	UK DOS	1986
Bendaja	Series G744	2339/II	US DMA	1967
Bolahun	Lib 50	2541/II	UK DOS	1984
Bong Town	Series G744	2538/IV	US DMA	1968
Cuttington	Lib 50	2739/II	UK DOS	1987
Daakoi	Series G744	2439/III	US DMA	1967
Dambala	Series G744	2339/III	US DMA	1967
Gbalatuah	Lib 50	2639/I	UK DOS	1987
Gbarnga (North)	Lib 50	2739/III	UK DOS	1987
Gbarnga (South)	Lib 50	2739/IV	UK DOS	1987
Gbatala	Lib 50	2638/I	UK DOS	1987
Jorwah	Lib 50	2739/I	UK DOS	1987
Kamatahun	Lib 50	2541/III	UK DOS	1983
Kle	Series G744	2438/III	US DMA	1968
Kolahun	Lib 50	2541/I	UK DOS	1983

Kongo (1)	Series G744	2339/I	US DMA	1967
Kpademai	Lib 50	2641/III	UK DOS	1984
Kpein	Lib 50	2739/II	UK DOS	1987
Lake Piso	Series G744	2338/III	US DMA	1967
Maana	Lib 50	2739/IV	UK DOS	1987
Madina	Series G744	2338/II	US DMA	1968
Mamaka	Series G744	2438/I	US DMA	1968
Mendekoma	Lib 50	2541/I	UK DOS	1983
Robertsport	Series G744	2398/IV	US DMA	1967
Vaahun	Lib 50	2441/II	UK DOS	1983
Vai Tieda	Series G744	2438/IV	US DMA	1966
Velazala	Lib 50	2641/IV	UK DOS	1983
Voinjama	Lib 50	2641/I	UK DOS	1986
Vonzuahun	Series G744	2338/I	US DMA	1967
Ziggida	Lib 50	2741/III	UK DOS	1986
Zowiento	Lib 50	2738/I	UK DOS	1987

Negatives for these 1:50,000 scale maps are prepared from 1:250,000 map originals. Copies are available from the USGS (http://www.usgs.gov/) or the NGA, the succeeding agency of the "Defense Mapping Agency (DMA)" (http://www.nga.mil/), or from the document delivery offices at AMRS, Inc. (Thttp://www.africaminerals.com/)

Balatengia	2639/III	USGS	1972
Bella Yella	2639/IV	USGS	1972
Belle Mbaloma	2640/III	USGS	1972
Bopolu	2539/II	USGS	1972
Domamana	2540/III	USGS	1972
Gelahun	2540/IV	USGS	1972
Gondalhun	2540/I	USGS	1972
Jawa Jei	2440/I	DMA	1973
Kongba (Kumgbo)	2440/II	DMA	1973
Konia	2640/I	USGS	1972
Lomboba	2640/IV	USGS	1972
Pala Kole	2539/II	USGS	1972
Salayia	2740/IV	USGS	1972
Soso Camp	2439/IV	DMA	1973
Takpoima	2439/II	DMA	1973
Tawalata	2539/IV	USGS	1972
Weasua	2439/I	DMA	1973
Womalu	2540/II	USGS	1972
Zelagai	2539/I	USGS	1972
Zolowo	2640/II	USGS	1972